THE NEW EDEN

THE NEW EDEN

WILDLIFE IN THE CITY, AND DISCOVERING OUR SHARED HOME

JC Niala

First published in Great Britain in 2026 by Gaia, an imprint of
Octopus Publishing Group Ltd
Carmelite House
50 Victoria Embankment
London EC4Y 0DZ
www.octopusbooks.co.uk

An Hachette UK Company
www.hachette.co.uk

The authorized representative in the EEA is Hachette Ireland,
8 Castlecourt Centre, Dublin 15, D15 XTP3, Ireland (email: info@hbgi.ie)

Text Copyright © JC Niala 2026

Unless stated otherwise, all poems are written by the author.

Distributed in the US by Hachette Book Group
1290 Avenue of the Americas, 4th and 5th Floors
New York, NY 10104

Distributed in Canada by Canadian Manda Group
664 Annette St., Toronto, Ontario, Canada M6S 2C8

All rights reserved. No part of this work may be reproduced or utilized in any form or by any means, electronic or mechanical, including photocopying, recording or by any information storage and retrieval system, without the prior written permission of the publisher.

JC Niala asserts the moral right to be identified as the author of this work.

ISBN: 978-1-85675-585-6
TPBK ISBN: 978-1-85675-586-3
eISBN: 978-1-85675-588-7

A CIP catalogue record for this book is available from the British Library.

Typeset in 10.5/16pt Farnham Text by Six Red Marbles UK, Thetford, Norfolk.

Printed and bound in Great Britain.

13 5 7 9 10 8 6 4 2

Commissioning Editor: Jessica Minocha
Senior Developmental Editor: Rachel Silverlight
Copy Editor: Susanne Hillen
Creative Director: Mel Four
Production Controller: Sarah Parry

This FSC® label means that materials used for the product have been responsibly sourced.

Tunza mazingira yakutunze.

Sustain the environment so it may sustain you.

Contents

Introduction		1
Chapter One	Border Crossers: Urban Animals and the Boundaries They Blur	9
Chapter Two	Edges and Enclosures: How Gardens Help Us Rethink Belonging	29
Chapter Three	Fear as Coexistence: Power in the Same Space	47
Chapter Four	Hawks, Ibises and Other Sky Neighbours	69
Chapter Five	Grown in the Soil: Cultivating in Crisis	87
Chapter Six	Held in the Soil: Plants, Memory and Resilience	105
Chapter Seven	Beyond Rewilding: How City Insects Teach Us to Sustain Life	123
Chapter Eight	Soft Civilizations: How Wetlands Shaped Urban Worlds	147
Chapter Nine	The Hoofprint Beneath the City: Learning from the Food Systems of the Past	167
Chapter Ten	The Long Game: Seeing the Wood and the Trees	183
Epilogue	A Living City: Ever Becoming	203

Notes	207
Bibliography	221
Index	231
Acknowledgements	245
About the Author	247

Introduction

I grew up somewhere between the wild and the city. On land that had not yet been fully 'tamed' but was beginning to show the hallmarks of so-called civilization. It was land where pastoralists had once roamed freely alongside their 'wild' animal neighbours and agriculturally inclined human counterparts. Land with layers of stories and struggles – land that was never truly settled. The 6,000 acres that Karen Blixen once tried and failed to farm for coffee had been appropriated from local peoples and divvied up by the British colonial government for European settlers. When Blixen returned to Denmark seeking treatment for syphilis, her legacy endured through the land's name, which became known thereafter as Karen. She later wrote of her experiences in a book that went on to become a Hollywood film, *Out of Africa*, starring Meryl Streep and Robert Redford.

Remy Martin went on to develop the land,* and many decades later Karen became firmly established as a Nairobi neighbourhood, complete

---

* Martin was a property developer – not to be confused with the brandy.

with schools, churches, essential amenities, restaurants, shopping malls and a variety of housing estates, many of which were gated. When I was growing up, from the mid-1970s to 1980s, it was seen as a last stretch of bush clinging to the edges of Nairobi's shirt tails. Our local shop was a colonial-style, red-roofed brick building that seemed to rise out of the dust, looking more like an out-of-place country dwelling than a modern-day supermarket. It was set within a scattering of low-rise buildings that included Gethin and Dawson's petrol and service station, a bank, a post office next to another far less popular supermarket, and the Horseman restaurant, all set against the backdrop of the iconic Ngong Hills.

Ngong, meaning 'knuckles', was the name given to the hills by Maasai peoples who had lived in the area for centuries. If you clench your hand the hills perfectly resemble the knuckles of your fist.

The story goes that, in the days when the earth was young, a giant roamed the land, terrorizing Maasai peoples by devouring their cattle. Fearing that once the cattle disappeared they would be next, Maasai peoples turned to their Laibon (spiritual leader), who in turn appealed to the Queen of the Termites.

Moved by their plight, the Queen promised to help. She assured them that when the giant returned, she would rally her termites to do whatever they could to stop him. Sure enough, the giant came back, rampaging through the land and gobbling up the last of the cattle. After his feast, he collapsed into a deep, satisfied sleep. That evening, the Queen kept her promise. As darkness fell the termites sprang into action, working tirelessly through the night to bury the giant. But as dawn broke, the giant was not completely buried, and only the back of his knuckles remained visible. Those knuckles, it is said, became the Ngong Hills we see today.

This story captures the complexities of Karen, a place where boundaries blur, and histories are both visible and hidden. It is a myth that explains a natural landmark: an echo of folklore that was shared among people who lived on this land long before Kenya's colonial era. It is an area whose original name was erased and replaced by a European one. A suburb that emerged from the heart of the savannah.

*

We had arrived – it seemed to me – in a magical place. My parents struck a deal with an architect who had begun building a house in Karen for a client who lacked the resources to finish it. As a result, we moved into an unfinished home with an equally unformed garden, a steep plot that sloped down to a stream, overgrown with tall *mabingobingo* grass, also known as elephant grass (*Cenchrus purpureus*).

I spent every moment that I could out in the back garden among the trees in the long grass. I was convinced that little people must live there, and I'd often go out to look for them. One day what I found instead, when I pulled back some grass, was a large and long snake. It was just there, nestled in the mabingobingo, silent and unmoving. It was only when I slowly raised my hand to reach out towards it that I saw a slight shiver, so I pulled my hand back so as not to disturb it. Its shiver made me pause. If *I* had been sleeping, and someone had reached out to poke me, I would not like it at all. The snake looked back at me as if it understood that I just wanted to look. It was a dark colour, and seemed to go on forever; certainly longer than my toddler height. We both stayed very still, as if acknowledging that it was nice to be hanging out at a respectful distance. When I did finally move away I did not do so in a hurry. I had enjoyed the snake's company, and something had passed between us – something peaceful and calm and in direct contrast to the mayhem that ensued when I told my mother about it later.

The snake had to be dealt with immediately. The mabingobingo was to be cut back. And I was to stay indoors. Oloo, the gardener, was called, and a conversation that began within my earshot continued out of it. I knew that something major was going to happen, but the full effects of how the garden would be rearranged would take several weeks to unfold. Eventually, the little people would have to find a new home as short and tidy Kikuyu grass (*Cenchrus clandestinus*) replaced the wilder mabingobingo. Once things had calmed down, my mother reassured me that the snake was no more, that the threat had gone. I failed to explain that I did not feel like that about the snake. We had shared a world just fine. The words did not come, and in any case would not have been heard. The grown-ups were busy shaping the natural world beyond our veranda to make it safe for me to be in. If the snake had been killed, it was my fault. I should not have said anything. Something in me registered that loss. The hush that followed the cutting of the mabingobingo never quite left me, and I think it's part of why I've listened to environments I live in ever since.

Something fundamental shifted for me in that moment. My mother, in her well-meaning panic, had enforced a boundary between me and nature, a boundary I hadn't even known existed. It seemed to me that both the snake and I 'belonged' in the mabingobingo. More than that, I began to wonder, why was no one considering the other perspective? What if, instead of the snake being the intruder, it was I who had entered the snake's world?*

✱

Nairobi is the only capital city in the world with a national park within its borders. Despite the park's modest size of 29,000 acres (117 square kilometres), it supports a diverse array of wildlife, from black rhinoceroses to lions. The park offers a striking landscape of open grassy plains dotted with acacia trees and scattered bushes, creating a classic feel of East African

---

* Pythons are sacred in my Luo community – one in particular called Omieri is known for bringing blessings and looking after our collective well-being. This is not the case for the Samia community that my mother comes from.

savannah. The backdrop of Nairobi's city skyline forms a unique contrast to the wild setting, blending nature and urban life. Its terrain includes rolling hills, small streams, and patches of dense forest along the banks of the Athi River. Seasonal wetlands attract a variety of birdlife, while the expansive plains serve as grazing grounds for zebras, giraffes and antelope. Both the animals and the welcoming climate attracted European settlers to this 'exotic' and yet highly habitable place.

Before British settlers arrived in the late 1800s, the Athi plains were home to thriving wildlife and an ethnically diverse mix of transitory settlements. The park's establishment reflects both colonial conservation ambitions and the displacement of local communities. With the arrival of colonial rule, white settlers claimed the fertile highland territory for their homes and farms. The impacts of European settlement around what is now Nairobi have been well-documented for two communities: Maasai and Kikũyũ peoples. The pastoralist Maasai coexisted with the region's wildlife, herding their cattle across the open plains, while Kikũyũ* peoples farmed the fertile, forested highlands situated above Nairobi. The European settlers displaced people from the nature they had long stewarded, and in ways that had simply not existed prior to this time. It's a place where people and wildlife have always lived side by side, sometimes in harmony, sometimes in conflict.

*

My parents taught me how to read a city such as Nairobi. Without maps or guidebooks, we would learn by observing how people connected with the natural world around them. I was shown how to pay attention to the small things: which trees lined the streets, how people gathered in parks,

---

* The spelling of Kikũyũ reflects elements of the original Gĩkũyũ language while keeping the terminology accessible to English readers.

or even how water moved through a city. My parents also made it a rule to always be respectful whenever we travelled to other places by learning the basics of the local language – how to say hello, please, thank you, and even how to count to ten. Looking back, it's probably no surprise I became an anthropologist. Those early lessons sparked a lifelong curiosity about the world and the constant interplay between people and nature.

Growing up in Nairobi, I saw this relationship unfold everywhere, and those early experiences taught me that the lines we draw between nature and human life are more blurred than we like to think. My fascination with cities and the natural world within them grew as I watched Nairobi transform around me. Take Karen, for example, where the old two-storey Horseman restaurant is now a modern multistorey building. The open fields where I used to ride horses have been fenced off, replaced by homes and business parks.

But it wasn't just what happened outside our compound that left an impression.* Inside our home was another kind of world – a place where people from all over Kenya and beyond would come and stay. Some came for a weekend, others for years. My parents, like their parents before them, kept an open house. Visitors brought with them stories of places I had never heard of, photographs of cities I could only dream of, and invitations to visit. I have been fortunate to travel to more than fifty countries and an endless number of cities.

Another place to which I feel deeply connected is London. My parents met and married there when they were students, something I have reflected on all my life, given that they come from markedly different parts of what was then the Colony of Kenya. When I was a child, my father would remind me, as we stood at Charing Cross, of how his colonial education had taught him to think of this notional centre of London as the heart of the

---

* A compound is an enclosed piece of land around a house or group of houses.

INTRODUCTION

British Empire. It was not until I was an adult that I appreciated the vast differences he would have had to navigate, from the fresh wetlands he grew up on to the smoggy London streets of the 1960s. My parents taught me to traverse cities, not through an A-to-Z or Google Maps, but through the ways in which nature and culture intersect. For me, London is a network of waterways, from the canals you can follow on foot from north-west London to the city's centre, to the Thames weaving through its heart. The city is a patchwork of markets, from Liverpool Street to Shepherd's Bush, and an ever-changing expanse of parks, from the grand stretches of Hyde Park to the wild beauty of Hampstead Heath. London, like many cities around the world, is full of nature – from tiny mosquitoes that live only in the London underground to whales that sometimes make their way along the Thames.

It's all there if we learn to pay attention.

This book is an invitation to reconsider our relationship with nature, especially in urban spaces. It's easy to think of cities as human domains, as places that belong to us alone. But the truth is that nature is always present, always finding ways to thrive, even in the most unexpected places. If we take the time to look, to really see, we might be surprised by how intertwined our lives are with the plants and animals around us. Cities are ecosystems too – complex, dynamic and full of life. Yet we often walk through them with blinkers on, unaware of the richness that exists just beyond our sight.

This book uses history, cultural commentary and the stories I grew up with to trace our relationship with the natural world in our shared spaces. It begins in the intimate terrain of memory and experience – moments that shaped my understanding of what it means to live with nature rather than simply in it. By describing the wide variety of more-than-human neighbours that we live with, from those that dispose of our waste to apex predators, this narrative considers what it means to live well alongside animals we might loathe or fear. It explores how communities create and care for green spaces, often in ways that challenge dominant narratives about what cities are for and who they serve. And it examines the broader systems – policy, history, language and design – that shape the possibilities for coexistence in the urban environment. Woven through each layer is a central invitation: to notice more closely, to care more deeply, and to imagine more courageously what kinds of lives – human and more-than-human – we might nurture in the places we call home.

Because, ultimately, this isn't just about noticing nature, it's about what we risk losing when we forget we are part of it. As cities grow, as landscapes shift, the danger is not only ecological but emotional. We lose memory, meaning, connection. We forget how to live well with other beings. If we want a future that is not only liveable but meaningful, we must learn to care differently. That begins by paying attention – to the cracks in pavements, the stories in trees, the silence that follows loss.

This book is my attempt to explore the relationship between nature and humanity and to invite you to look at cities with new eyes. It is an invitation to marvel at the resilience and adaptability of nature, even in the most urbanized environments. It is a call to reconnect with the wonder that comes from recognizing the presence of life all around us, in places we might not expect. From the weeds growing in pavement cracks to the birds nesting in high-rise buildings, nature is always there, quietly insisting on its place in the world.

As you read, I hope you'll be inspired to look more closely at the urban spaces you inhabit. There's a richness there, a beauty in the way life persists and thrives despite our best efforts to control it. And perhaps, in doing so, you'll find that the question, 'What's it doing here?' becomes less about intrusion and more about belonging – about recognizing that we are all, in our own ways, just trying to find our place in the world.

Chapter One

Border Crossers: Urban Animals and the Boundaries They Blur

I hesitate before getting into the taxi. The driver is easily six and a half feet tall, burly, missing a few teeth, and his tattoos run like stories down his arms. His long dark hair is pulled back into a messy ponytail. But after 45 minutes of waiting, as taxis rushed past me despite my increasingly desperate waving, and with the characteristic San Francisco drizzle beginning to seep through my clothes, I climb in.

I ask him to take me to a bar near City Lights bookshop. The thought of returning to my lodgings – thin-walled and noisy, filled with the unsettling dramas of strangers – makes me shudder. After a long day in an office where everyone discussed evening plans that pointedly did not include me, the bustling anonymity of Polk Street's nightlife isn't the kind of soundtrack that will assuage my loneliness.

Our eyes meet briefly in the rear-view mirror. Without thinking I blurt out, 'No taxis would stop for me.'

'That's because you're black,' he replies, evenly.

A moment passes. A black woman and a white man size each other up, each of us confronting instant judgements – and then we burst into laughter. When we reach the bar, he refuses my money with a gentle shake of his head.

'We've all had that kind of day.'

I step onto the street feeling light enough to pause, noticing a pigeon gliding down to join another, sheltering under a narrow ledge from the drizzle. There's something quietly reassuring in their instinctive companionship, finding solace together in the midst of the city's indifference.

Inside, I order a drink at the bar and an elderly man moves to sit beside me. Normally I'd recoil, retreating into guarded solitude, but tonight, buoyed by the unexpected kindness from the taxi driver and perhaps even by the quiet pigeons outside, I don't shift away.

He tips his black trilby forward slightly, introducing himself as Tony. Surprising myself, I respond with my real name. Loneliness has a subtle scent. It's the passenger who sits beside you a little too cheerfully on the bus, sharing their soul in more detail than you're comfortable hearing. Often, we pull away, fearing to see ourselves reflected back. Two lonely people who might otherwise have connected end up isolated by a mutual wariness. Tony and I pick up that scent on each other now. Tonight, instead of retreating, I remain open, quietly waiting to hear what this stranger might say.

It turns out that Tony once lived with a pigeon. He says this matter-of-factly, as though it's a common arrangement. Initially unconvinced, I glance at the bartender, who meets my gaze with an amused nod. Tony's pigeon, apparently, is a known quantity around here.

Tony found her on the sidewalk outside his apartment, sick, feathers in disarray, utterly vulnerable. Without much thought, he took her in. With patience and care, he nursed her back to health, though he insists that she healed him just as much as the other way round. Over weeks and then months, the pigeon grew stronger, healthier, never once showing any desire to leave. Even with windows wide open, offering her the sky, she stayed, hopping behind Tony from room to room, a steadfast companion in his solitary life.

As Tony speaks, affection softens his weathered face. His silent gratitude is palpable, reserved for the unexpected presences in our lives, those brief yet profound companions reminding us that we're not entirely alone.

'Then this morning,' Tony continues, 'after breakfast, she just left. Flew straight out the open window.' He smiles, halfway between sadness and amusement. 'Left me a little farewell present on the table, too. Her way of saying goodbye, I guess.'

We laugh softly, sharing the space left by the absent bird.

Tony's pigeon reminds me of my mother's long-forgotten pigeons, and so I share my story with him in return.

When I was a child, Uncle Aden frequently visited our home, often arriving in a car packed with as many relatives as he could, just about, legally squeeze in; aunties and children, all of them tumbling out in a lively chaos that always delighted my mother. Among the surprises that Uncle Aden regularly brought, one particular gift stood out: a small flock of homing pigeons.

My mother, who generally approved of animals only if they were either useful or strikingly beautiful, made a rare exception for birds. She adored them all. Uncle Aden, fully aware of this, convinced her to let our gardener, Oloo, construct a carefully designed pigeon loft behind the cowshed. On a lazy afternoon, from my usual perch on the veranda, I overheard snippets of their conversation – Uncle Aden passionately instructing Oloo on the finer points of pigeon care, training and homing instincts.

My mother's voice, cautious yet hopeful, carried across the garden: 'Are you sure they'll come back?'

'Of course,' Uncle Aden assured her with unwavering confidence. Oloo nodded along enthusiastically, his smile suggesting either total comprehension or polite bafflement – I could never quite tell which.

Weeks later, my family gathered expectantly as the pigeons took their first flight from their new loft. We stood watching, eyes wide with excitement, hearts buoyed by Uncle Aden's optimistic predictions.

It was clear, by dusk, with the sun slowly dipping, that the pigeons weren't coming back. I remained outside long after everyone else had retreated indoors, staring at the empty loft as though sheer hope could summon them home. Inside, my mother was already on the phone to Uncle Aden, who swiftly promised another flock. But this second group vanished just as decisively as the first, leaving the carefully built loft empty once again. Eventually, Oloo dismantled the pigeon house, quietly repurposing its wood and wire elsewhere.

Even as a child, pigeons stirred conflicting emotions within me. Part of me yearned for the thrill pigeon fanciers speak of, the joy of watching beloved birds soar confidently into the sky, then faithfully return home. Yet another part of me cheered silently at their defiance, their fierce autonomy. The birds chose freedom over comfort, adventure over predictability. Whenever I spot pigeons in city streets these memories resurface – reminders that even in the most human-centric spaces, wildness persists, and animals retain the power to chart their own paths, subtly redefining what it means to adapt to urban life. As I finish relating this to Tony he smiles knowingly. 'Seems like pigeons have their own ideas about home,' he remarks gently.

Sitting here, sharing these stories, I realize that both my mother's pigeons and Tony's quietly unsettle the categories we like to rely on: wild and domestic, pest and pet, intruder and guest. They remind me that animals we dismiss as background noise in our cities are often making complex decisions about safety, survival, even affection. We might recognize them as adaptable, but maybe what's really confronting is their refusal to be contained by our expectations. If pigeons can define where and how to belong, then perhaps we need to reconsider what urban belonging looks like. It might not always be about fitting in, but rather about finding ways to stay on your own terms. That realization takes me back further still, to when I was five and sensed for the first time that belonging wasn't a simple matter of who was invited but of who insisted on being present.

My parents had agreed to my having a cat because they hoped it would make a good indoor playmate and keep me away from the allergies that flared up when I was in the garden. Hopscotch, a grey tabby, was my first true pet, in as much as cats ever can be such a thing. She was an extraordinary hunter, a skill that endeared her to my otherwise pragmatic mother, who values animals only if they 'earn their keep'. For my mother, this didn't just mean being useful in a practical sense – by controlling pests, say – but could include simply being beautiful. Hopscotch fulfilled both criteria, excelling as a hunter while possessing the charm of a sleek and striking tabby. Sadly, she only lived with us until it became clear that I was allergic to her too. Even so, she left a lasting impression on me, not just for her companionship but for the role she played in my early understanding of how animals fit into human spaces.

When I was five, Hopscotch and I were so close that she used to sit on my shoulder, a bit like an old-fashioned pirate's parrot, whether I was sitting reading or even if I was wandering around the house. And we were indeed playmates. Alongside the toys my parents bought for us, from the time she was a kitten I loved making my own – mostly versions of wool strings with small balls or other objects attached. I delighted in teasing her, dragging these along the floor and then snatching them away just as she pounced, watching with endless amusement her determined attempts to catch them. And she had her own way of 'calling' me to play. If I was reading at the dining table, for instance, she would weave around my legs and let out a short, higher-pitched purr – her signal that she was in the mood for a game. One day, thus called, I followed her into the sitting room, expecting to find a toy she had retrieved. Instead, I was startled to discover a group of six baby white mice, which Hopscotch proudly presented to me.

I remember the moment with vivid clarity. As soon as I saw the tiny white pups, I understood: Hopscotch had killed their mother. That meant that no one would be feeding them, and so of course they must be hungry. I tried to reason it out the best way I knew how. Mice ate human food – that much I'd learned from the way we kept them out of our pantry. But these weren't full-grown mice. They were babies. And babies, in my limited experience, needed something different. Human babies drank uji (traditional East African breakfast porridge made from millet, sorghum or maize flour, fermented or unfermented), but mice seemed to prefer raw things. So, I thought, maybe *wimbi* (finger millet, *Elusine coracana*) would be perfect – still soft, but more natural. It made perfect sense to me at the time. I was determined to save them. I had a plan.

First, I needed a home for them – a cardboard box would do, and if I poked some holes in it they'd be able to breathe. Then they would need somewhere soft to sleep in the box. *Pamba* (cotton wool) would be ideal. I'd turn one corner into a bedroom, and another into a little dining room. I felt very proud. They would be warm. They would be safe. They would have everything they needed. I would take care of them.

I also knew two critical things as I sought to protect the mice: first, if my mother discovered them their fate would be sealed – she would regard them as pests and dispose of them without hesitation. Second, Hopscotch couldn't be trusted, given her role in how they had come to be in my charge. Determined to look after them, I hid the pups in what we called the attic, a chaotic space filled with odds and ends that my mother believed might one day prove useful. It seemed the safest place to keep them away from both my mother and Hopscotch.

Keeping the pups a secret soon proved to be a monumental task for a child of my age – not because I was tempted to tell anyone (I relished the thought of them as my special, secret friends), but because the demands of a five-year-old's life left little room for clandestine caregiving. Mealtimes, bath times, and supervised playtime constantly pulled me away from my mission. Asking an adult for help was out of the question, knowing that any hint of my new friends would make its way back to my mother.

Despite my best efforts (and because what I was feeding them was entirely unsuitable), within just a couple of days the pups died. When my mother eventually found them, she assumed that Hopscotch had simply finished what she'd started, and my beloved cat was rewarded with special treats for her 'efforts'. I said nothing at the time, but those tiny, helpless pups lingered in my thoughts for years. Only much later did I realize that they were probably albino rats (*Rattus norvegicus*) – rare in the wild but common in urban settings, bred specifically for laboratories. Reflecting on this, it struck me as deeply ironic: we often tolerate – even encourage – rats in our cities, provided they exist within tightly controlled environments. Inside labs, we dictate how rats live, and, inevitably, how they die. Yet the moment these animals step outside those boundaries and into spaces we call our own, they are quickly dismissed as vermin – pests to be exterminated without thought or care.

Looking back, it seems that those tiny albino pups, doomed though they were, offered me my first real glimpse into the lives of animals we usually ignore. Even as they clung to life in the attic of a suburban home, their presence forced me to question the tidy lines we draw between what belongs and what doesn't. In cities around the world, creatures like rats, pigeons and foxes don't just survive, they adjust, innovate and endure. Their genius lies in being constantly overlooked, in quietly reading the urban landscape and bending it to their needs. I didn't know it then, but those pups were early guides into a deeper truth: our cities are not solely human creations.

They are shared, shifting ecosystems shaped as much by animals as by our own designs.

*

Manuel Berdoy, an award-winning zoologist at the University of Oxford, became curious about the behaviour of lab rats – animals meticulously bred for life in artificial environments. It led him to conduct an experiment that challenges our understanding of what it means to be wild or tame. He took 75 rats, whose ancestors had been born and raised in cages without any exposure to the natural world for 96 years (or about two hundred generations), and released them into a farmyard to see what would happen. The rats were moved from a sterile environment of plastic cages with artificial lighting, temperature control and a steady supply of nutritionally balanced pellets to an open space the size of a tennis court – with grass, stones, straw bales and hurdles like ladders. Though lab pellets were available in the new environment, they quickly expanded their diet, trying a wide range of food from fruits to eggs.[1]

The remarkable result was that, instead of descending into chaos, the rats began to form social hierarchies and quickly established well-worn paths, known as 'runs', which criss-crossed the farmyard. These were the routes they used repeatedly to move between shelter, food and nesting areas, a clear sign of organization and adaptation within just a few days of arriving in their new home. Despite the fact that lab rats are bred to be more docile, it turned out that they hadn't lost their adventurous spirit and they displayed an uncanny ability to adapt to what was for them a new more natural environment.[2] They quickly re-learned how to forage, defend territories and even avoid predators – behaviours thought to be far removed from their laboratory lineage. This experiment suggests that natural instincts deeply ingrained from millennia remain just beneath the surface, ready to emerge when circumstances demand.[3]

Berdoy's study reveals just how blurred the boundaries between the domesticated and the wild truly are. Even in cities, where we often assume that animals are either fully tamed or entirely feral they are shifting between roles, environments and behaviours in ways that defy neat categories. Rats, pigeons, foxes and countless others navigate our urban ecosystems with remarkable ingenuity. They aren't 'city animals' by nature but creatures responding to a specific environment – one defined by us yet shared with them. In another setting, they would behave entirely differently, just as humans would. A rat scuttling through the London underground or nesting in a research lab isn't fundamentally separate from its forest-dwelling relatives; it's simply living in a different context, proving its ability to adapt. But our perception of these creatures reveals our ability to underestimate them.

It remains a fascinating paradox that, despite being taught that these creatures are vermin, they continue to exert a hold on our imagination, and a sympathetic one at that. In popular culture and in folk tales, mice are cast as loveable domestic and often rural creatures, who are just doing their best to survive in a big and intimidating world.

It's no coincidence that I found myself rooting for Jerry in *Tom and Jerry*. Although I was enamoured with cats, it was Jerry's cuteness and inventiveness that earned a place in my heart, something that I think stemmed largely from the near-universal appeal of support for the underdog. This was just the kind of world a small child could imagine, but also the kind that adults could retreat to when everything seemed to be going against them.

A classic example of this type of underdog is Fievel Mousekewitz, a young Russian mouse who stars in the 1986 animated musical *An American Tail*. Separated from his family after immigrating to the US, Fievel must navigate a new and dangerous world in the hope of being reunited with them. Similarly, in the 2008 film *The Tale of Despereaux*, a heroic mouse takes on the role of rescuer, confronting a villainous rat named Chiaroscuro who has imprisoned the human Princess Pea. Despereaux is an outsider, not just because he has lost his mother, but because he doesn't quite fit in with the other mice. E B White understood this deep-seated empathy for the underdog and took it one step further with *Stuart Little*, a character who isn't just mouse-like but a mouse-child hybrid: delicate, vulnerable and determined. By blending human traits with the mouse's innate charm, White tapped into our impulse to root for the overlooked. Small, determined and endlessly creative, mice effortlessly invite identification – especially when the odds are stacked against them.

In the ancient world, mice weren't always seen as loveable. In fact, their association with danger and power runs deep. One of the more surprising figures connected to mice is Apollo, one of the most multifaceted gods in Greek mythology. Known as the god of light, prophecy and order, Apollo was also linked to both disease and healing – and, curiously, to mice. In Book I of *The Iliad*, the priest Chryses appeals to Apollo to punish the Greeks after Agamemnon abducts his daughter. Apollo responds by unleashing a devastating plague, an act that sets the central conflict of the epic in motion. Because mice were understood at the time to be harbingers of disease, Apollo earned the epithet *Smintheus*, or 'Lord of Mice', embodying both destruction and cure.[4] This duality was immortalized in temples, statues and coins – some of which depict Apollo with a mouse near his hand or foot. The ruins of one such temple still stand in the village of Gülpınar, on the Biga Peninsula in Turkey.[5] These ancient associations remind us just how emotionally charged and culturally embedded our relationships with mice have always been – swinging between reverence and revulsion, utility and fear.

Yet not all rodents have been granted such symbolic complexity. If mice have been mythologized, rats have been vilified. Unlike their smaller cousins, rats remain burdened by a reputation shaped not by storybooks or symbols but by centuries of fear and disease. The bubonic plague, often referred to as the Black Death, was one of the most devastating pandemics in human history, peaking in Europe between 1347 and 1351. The name comes from the darkened patches that appeared on the skin of those afflicted. Caused by the bacterium *Yersinia pestis*, the disease spread rapidly, killing an estimated 25-50 million people, which was roughly a third of Europe's population at the time. Symptoms included swollen lymph nodes, fever, chills and, if untreated, a high likelihood of death. The consequences were far-reaching, triggering labour shortages, economic collapse and massive shifts in Europe's social fabric.

For generations, it was understood - and taught - that the plague had been spread primarily by fleas that lived on infected rodents, especially black rats (*Rattus rattus*), who thrived in the close quarters of medieval cities. These rats, living in such proximity to humans, were believed to act as disease reservoirs, enabling transmission. The historical records focused on human experiences, so although rats could be infected, it remains unclear how lethal the bacteria was to them or to other species. Still, while rodents as a group - rats, mice, squirrels - were implicated, it was rats who became the primary symbol of pestilence.[6] Plague-spreading mice rarely appeared in the popular imagination.

As the research has deepened, it turns out that the correlation of rats with the problem of the plague is complex. The unsanitary conditions of medieval life also played a critical part in the disease's spread. And, despite its profound impact, the exact mechanisms of plague transmission are still not clear. Historical and archaeological evidence supporting the claim that rats and their fleas were the primary vectors of the disease is increasingly being challenged. Recent research suggests that human ectoparasites, such as body lice and fleas, may have played a more significant role in the spread of such epidemics in pre-industrial Europe.[7] This evolving understanding of the plague forces us to reconsider long-held assumptions about rats and their role in historical urban environments.[8] While they've long been cast as villains in the narrative of disease and decay, closer scrutiny reveals a more nuanced picture.

*

One of the clearest embodiments of the tension between humans and rats is the figure of the rat-catcher – a person tasked with outsmarting an animal that seems, in many ways, to mirror our own intelligence and determination. At the heart of this conflict is a battle over both space and food. In rural areas, rats consume crops; in cities, they thrive on the excesses of human life – our wastefulness, poor sanitation and decaying infrastructure. The abundance of discarded food in urban areas does more than attract rats, it allows them to flourish. Once established, they are notoriously difficult to remove. Their intelligence, adaptability and caution, particularly around new objects like traps, often frustrate even the most seasoned exterminators. We may call them pests, but rats reflect us back to ourselves in uncomfortable ways: opportunistic, resourceful and always alert to danger.[9]

City exterminators, who often express mixed feelings about rats (not enough of them and they'd be out of a job), note how rats' behaviours echo our own lives. Writer Robert Sullivan spent a year studying rats and those who deal with them in New York City, documenting the resilience of these creatures and their complex social structures. Rat-catchers, on the one hand, take pride in their ability to outwit rats, but they also grapple with the ethical implications of extinguishing such resourceful lives.[10]

I stumbled upon the long history of rat catching almost by accident, when my friend Daisy was, as she put it, being terrorized by a rat in her house. She had tried everything short of poison, but she was consistently outmanoeuvred by the rat. Exasperated, she was on the verge of using poison when I suggested she borrow our family dog, Bear, a dachshund/

Jack Russell terrier mix with a clear hunting instinct. Daisy later recounted how she'd never seen two creatures move so fast. In a blur of movement, Bear chased the rat out of the house and to its demise. Following this adventure, Bear took quite a liking to Daisy. Although I would have preferred the rat to have been relocated rather than killed, the episode sparked my curiosity about rat-catching methods and the ways in which humans have interacted with rats in urban spaces throughout history.

*

It turns out that Victorian England also struggled with rats, and, much like Daisy, people weren't just concerned about the diseases the animals could spread. The very sight of them often caused mental distress. Terrier dogs, like our beloved Bear, became highly valued for their rat-dispelling abilities. Rat catching even became a profession, providing livelihoods for many. The most notorious rat-catcher of the era was Jack Black, who styled himself as 'Rat-Catcher to Her Majesty The Queen'. Based in Battersea, London, Jack was as much a showman as an exterminator, dressing in flamboyant outfits adorned with a leather sash decorated with cast-iron rats. Notably, he didn't kill all the rats he caught. Any with unusual colouring were spared, bred and sold as pets, reportedly even to Queen Victoria herself.[11]

This shift in perspective, from seeing rats not simply as threats but as fellow urban inhabitants, has surfaced in unexpected places, including art. The graffiti artist Banksy, ever attuned to the overlooked and subversive, has long used rats as stand-ins for the urban everyman: scrappy, resilient and constantly adapting. In his 2005 exhibition *Crude Oils*, which involved live rats, the message was clear: the animals we dismiss as pests may be some of the most successful cohabitants of our cities. Banksy's rats are both symbols of rebellion and a reminder that adaptability itself is a form of genius, quietly thriving beneath our feet.

THE NEW EDEN

Taking his cue from the enduring myth that in London you're never more than six feet away from a rat, Banksy's *Crude Oils*, staged in London's Notting Hill, featured around 150 live rats roaming freely among the artworks, which included boldly reimagined classic paintings and sculptures. One striking installation, which involved the rats darting in and out of a human skeleton posed as a gallery attendant, created a particularly unsettling juxtaposition. The presence of the rats heightened the atmosphere of decay and disruption, transporting visitors into a dystopian vision of urban life where the boundaries between art, life and decomposition collapsed. Only rats running across sandalled feet could create that sort of visceral reaction.[12]

The alteration in our attitude to rats can also be seen in popular culture, such as in the film *Ratatouille* (2007). This animated film cleverly subverts the traditional image of rats as filthy, unwanted creatures by presenting Remy, a rat with refined tastes and culinary dreams, as its hero. Instead of being a signifier of disease or decay, Remy is depicted as intelligent, creative and, when given the chance, capable of extraordinary achievements. His journey, in which he overcomes human prejudice in order to pursue his passion, could be said to mirror the cultural shift towards recognizing the complexity and adaptability of rats.

As well as challenging the long-standing stereotype of rats as villains, this portrayal invites us to reconsider our relationships with animals often dismissed as vermin. By placing a rat in a setting as inherently 'human' as a gourmet kitchen, *Ratatouille* bridges the divide between human and animal worlds, highlighting the unexpected connections we share.

One such connection has been found in the greening of urban spaces. Greening cities benefits both humans and rats, though in ways that might initially seem paradoxical. Greener cities such as Eindhoven in the Netherlands often host larger rat populations, but these rodents are less visible (and as such less mentally distressing for human beings), living in undergrowth and other natural habitats. They quietly support urban ecosystems by cleaning up waste and serving as prey for predators, enriching biodiversity. At the time of writing it is too early to draw definitive conclusions, but studies in Dutch cities have laid the groundwork for further research.[13] These findings suggest that a balance is possible, whereby, as cities grow greener and human behaviours encourage rat abundance, we might find new ways to live alongside these creatures who we often malign, even as they have the power to improve our lives.

I think again of Tony and his pigeon, and the way his whole face softened when he described her loyalty, her presence, the unspoken comfort of her just being there. She wasn't exotic or rare. She was, by most measures, vermin. But she gave his life a kind of rhythm, a thread of companionship in the city's often anonymous sprawl. And perhaps that's what these overlooked animals do best, not so much offer something back as interrupt the scripts that we have written about who belongs where and why. A flash of wings, a rustle underfoot, and suddenly the world feels less sealed off. They force us to live with contradiction and to make room for the wild in the most controlled environments. Tony did more than rescue a pigeon. He invited nature back in. And maybe, without quite realizing it, she reminded him that not all connection needs explanation.

In previous generations, this kind of closeness wasn't unusual. Animals were woven into the texture of everyday life. Pigeons, in particular, weren't merely tolerated; they were kept, trained, relied upon. People bred them, housed them and carried them across borders and battlefields. Their presence in cities was often intentional. They lived in our attics, our sheds, our eaves. And perhaps that long history of proximity is why, even now, we respond to them in ways that are more complex than we realize. We've shared space with pigeons for so long that, whether we acknowledge it or not, they feel familiar, straddling that strange line between pet and wild creature.

In wartime, pigeons were nothing short of heroes. During both world wars, they carried crucial messages through dangerous conditions, often saving lives. One of the most famous is Cher Ami (*Columba livia domestica*) who was one of 600 homing pigeons used by the US Army Signal Corps in France during the First World War to provide crucial battlefield communication. Homing pigeons possess an extraordinary ability to navigate back to their home lofts, even across unfamiliar terrains. They do this by using low frequency sounds to create a 'sound map' in their minds.[14] Messages were written on small notes and placed in canisters attached to the pigeons' legs, allowing them to carry vital information across battle zones as telephone and radio communications were still unreliable at that time.

Cher Ami carried out his heroic service during the Meuse–Argonne offensive in October 1918. Serving with the 77th Division, also known as the 'Lost Battalion', Cher Ami faced his most dangerous mission. The battalion had been cut off behind enemy lines, enduring relentless bombardment by the Germans who had also shot down all other pigeons. The battalion was running out of options to make their position known. Their dire situation left them with only one pigeon – Cher Ami.

Under heavy German fire, Cher Ami carried a message from Major Charles Whittlesey that saved 194 men from friendly fire. Despite being shot through the chest, losing a leg and becoming blind in one eye, he flew 25 miles in under 30 minutes to deliver the crucial note. Cher Ami's bravery earned him the Croix de Guerre from the French government.[15] Other pigeons were employed by postal services around the world, ferrying letters and tiny packages before modern communication methods replaced them. For centuries, these birds served as a bridge, connecting people separated by geography and conflict.

※

Given their significance, I was devastated when Ken Livingstone, London's first elected mayor, declared his war on pigeons.[16] To me, they had always been as much a part of London as its rodents, a fixture of the city's vibrant and messy ecosystem. At the time I was working on the Strand – a historic and bustling street in central London, known for its unique blend of culture, commerce and history. Running parallel to the north bank of the River Thames, it stretches from Trafalgar Square to Temple Bar, acting as a gateway to some of the city's most iconic landmarks. My lunchtimes were often spent either 'towards the water', sitting on a bench on a quiet section of the Embankment, or, on warmer days, in Trafalgar Square. There, I would watch tourists, feed the pigeons and think about my parents as I looked at Charing Cross.

Trafalgar Square, in particular, evoked the blurring of boundaries that defined so much of my childhood. I would stand very still in the square, arm outstretched with birdseed, imagining myself as the Bird Woman from the classic children's film *Mary Poppins* (1964), hearing the song 'Feed the birds, tuppence a bag' playing in my head as I waited for the pigeons to come. Often, my pigeon feed ended up in the hands of young tourists, much to their delight (and sometimes to their parents' dismay). There was something about the simple act of feeding pigeons that made the city feel softer, more connected. In that sense, I was not unlike Phil Daniels, who in 'Parklife', Blur's 1994 Britpop song, talks about how good he feels about feeding pigeons and sparrows.

And then came Livingstone, a man who lived just a few stops away from me on the Tube line, threatening to bring it all to an end. His campaign to banish the pigeons felt personal, a dismantling of one of the rituals that tethered me to the city. I remember the day I saw him on the platform as I headed into work. Before I could think, I called out, 'Ken – leave the pigeons alone!' He turned without hesitation and shouted back, 'They spread disease – flying rats!' before stepping onto his train and disappearing. I was so startled by the exchange that I forgot to get on the train myself and had to wait for the next one.

Livingstone's retort stayed with me, both for its brusqueness and for how casually it reduced these creatures to nothing more than carriers of disease. It was the same narrative we use to dismiss so many urban animals, as if their existence is only justified by their utility or lack of inconvenience to us. Yet for me, feeding the pigeons had never been about utility – it was about the way in which both pigeons and people had adapted to the city, and those small, almost imperceptible moments where the borders between human and animal dissolve, even in a city as bustling and fragmented as London. What, I began to wonder, could we learn if we stopped seeing pigeons – and by extension, all urban wildlife – as problems to be solved?

Reflecting on those tiny, helpless pups I had tried and failed to save, I keep circling back to a more unsettling question: who truly belongs in the city? The thing that unnerves us about urban wildlife is not just the risk of mess or disease – it's the quiet collapse of boundaries we thought were firm. Between wild and tame. Nature and civilization. These animals don't seek permission. They appear, endure and often thrive in spaces we assume to be ours. Cities are supposed to be human-made, human-run. But creatures like rats, pigeons and foxes slip through the cracks, chart their own paths, and build lives from what we discard. They do more than adapt to urban life – they expose how unwilling we are to share it.

At the time, I saw the rats simply as vulnerable. But now I see them as something more complex – messengers, in a way. Their presence delivers a truth we often choose to ignore: that not all lives are welcomed equally in urban spaces. By labelling certain animals as vermin, we erase their stories, their agency, their astonishing capacity to adapt. And that act of erasure isn't just something we reserve for animals. We do it to people, too.

In cities around the world, people too are told – implicitly or explicitly – that they don't belong. That they are out of place. That their presence is inconvenient, unsightly or disruptive. This sense of exclusion echoes the logic we apply to rats and pigeons; a logic that says that some lives are compatible with city life while others are not.

But these animals teach us about a quiet form of resistance. A way of saying: *I will live anyway*. That's what Tony's pigeon did. That's what the lab rats in Berdoy's study did. And that's what the rats we try to forget, erase or poison still do. They persist. They take up space.

> *In the city's bones*
> *Claws lay claim to what we waste*
> *Rats refuse to die.*

In these brief lines, rats become survivors quietly retrieving space from a world designed to exclude them. They remind us to reconsider not only the animal we label unworthy but also the people pushed to societies margins. Those living at the edges of recognition. Those whose presence feels too noisy, too raw, too complicated for the kinds of clean, curated cities we imagine we want.

If we start to take seriously the idea that more-than-human life deserves a stake in our urban future, then what does that mean for how we design and defend the green spaces within it? It begins, perhaps, in the overlooked cracks of the pavement – where even the most unloved lives insist on being seen. And from there, it reaches towards gardens – where we will turn next – and the ongoing fight over who has the right to thrive, to breathe, to belong in the spaces we share.

Chapter Two

Edges and Enclosures: How Gardens Help Us Rethink Belonging

My sleep is disrupted as I turn over to escape the light. It doesn't shift. Through the fog of jet lag, I hear my mother's voice.

'You forgot to draw the curtains,' she murmurs, barely awake.

I drag myself out of bed. This, I realize, is why they call Japan the land of the rising sun. Before I shut it out, I let myself look. From our room high up in the Akasaka Prince Hotel, Tokyo stretches out in every direction – all buildings and blossom. The sun now safely tucked away from my mother, she quickly returns to sleep. But I'm fully awake. It's ten past five in the morning and the city is calling.

This is our routine. My mother works in international finance, and I, though just a teenager, travel with her. We arrive in a new city, she heads to meetings, and I head out to explore. In the evenings, I report back with what I've discovered. When teachers complain she insists that I'm still getting an education – just not the classroom kind.

But I know that she prefers me to wait until we've properly arrived and are less bleary-eyed with jet lag. Her meetings don't start for a couple of days. I can't wait that long.

I gather my clothes, take the briefest of washes, and dress quickly. I'll be gone half an hour, back before she wakes up. Slipping down to the lobby, I ask the curious receptionist – clearly at the end of his night shift – what I should do while I'm in Japan. He pulls out a map, gestures animatedly, and explains what he calls the 'real Japan'.

Soon I'm outside, the morning bright and crisp. I make note of landmarks, committing them to memory, my own breadcrumb trail. Just near the station, something stops me. A park. Trees draped in blossom I've never seen before – not quite cotton wool, not quite confetti. Some petals are blush white, like pale socks tinged by something red left in the wash, others the soft pink of baby celebrations.

The park swallows me whole.

By the time it releases me, I've stayed longer than planned. I rush back to the hotel, the once-whizzy lift now achingly slow. I dash to our room and spot a thin strip of light under the door.

Damn.

I open it gingerly. My mother is upright in bed, very much awake. She opens her mouth –

'We're going to an authentic Japanese garden in Kyoto,' I say, before she can speak.

*

Maybe it was the bullet train that ruined me. I'd never experienced anything like it, leaning against the window, watching the world blur past in streaks of watercolour. Trains in the UK simply didn't move like that. Or maybe it was because I was still in my teens, and whatever illusions I had about being a worldly traveller evaporated the moment I found myself staring at 15 lumps of rock arranged on patches of moss, surrounded by pebbles, bounded by clay walls. The dry landscape garden is in front of the monastery at Ryoanji Temple. A famous Zen Buddhist monastery. Real Japan. Visitors look at it from the wooden veranda. Was this the famous garden? I'd convinced my mother that we were going to come here, and now we were committed for the rest of the day. She was thrilled – serene, even – as she sat perfectly still, gazing out at the rocks with her eyes half-glazed in what I assumed was spiritual rapture. Meanwhile, I was stuck trying to meditate on the meaning of life.

My early attempts to engage my mother in conversation – mostly along the lines of, 'how is this even a garden if it doesn't have any plants?' – were swiftly silenced by sharp looks from her and a chorus of international visitors eager to commune, reverently, with the rocks. I was less reverent. I couldn't understand why anyone would travel halfway around the world – or spend what seemed like a fortune – to sit in front of what looked, to my teenage mind, like an extremely tidy gravel pit.

An older American man sitting nearby must have picked up on my restlessness. He leaned towards me and asked, in a low voice, what I thought the rocks meant. I dithered, not sure whether it was a trick question or a genuine one, and mumbled that I had no idea. He smiled, not unkindly, and gestured at the spread of stones before us. 'Look around,' he said, motioning to the rows of visitors beside us, people of every age, from all over the world, cameras tucked away, sitting respectfully quiet. 'Maybe the rocks are the continents,' he continued, nodding at the specifically placed arrangements on the gravel sea. 'And the pebbles are the oceans that connect us all. We think we're separate, but maybe this is a reminder that we all share a world.' I didn't know how to respond. I wasn't ready to agree with him, but I wasn't ready to dismiss him either. His words lodged themselves in my mind, somewhere between annoyance and wonder.

*

Years later, when I revisited that moment through the lens of my adult work, I finally understood what was happening to me on that Kyoto temple veranda. What I'd experienced wasn't just teenage resignation, it was a cognitive shift. Underneath my initial fidgeting, I had noticed a certain calm begin to settle over me. Even though I had been determined to be annoyed, the longer I had stared at the rocks, the harder it became to hold on to that frustration. It kept slipping away from me like water through my fingers.

Today, psychologists and neuroscientists talk about certain types of natural settings as 'cognitive sanctuaries', places that allow the brain to move from directed attention – the kind we use to focus, plan and get through our to-do lists – to what's called 'effortless attention'. In such moments, our minds aren't blank, but gently held. We are engaged, but not taxed. The Zen rock garden, despite – or rather because of – its minimalism, offered exactly this kind of space.

Unlike a wild forest or a flower-filled park, which offer their own particular kind of benefits, the carefully composed simplicity of the *karesansui* garden bypasses sensory overload. The raked gravel, the unchanging rocks, the silence broken only by the sound of wind or birds – all these elements lull the nervous system into an easier rhythm. What I'd interpreted as boredom was, in hindsight, something closer to restoration. I had no name for it then, but what I felt was my internal noise beginning to dial down, giving way to something stiller. My mind was resisting, but my body had started to soften. It turns out that I had not imagined that 'gentle shift', something physiological was going on.[1]

What still intrigues me is how such spaces can be both deeply rational and entirely mysterious. The patterns of sand and gravel that soothe the brain can be charted in MRI studies. The visual harmony of asymmetrical design has measurable effects on blood pressure and emotional regulation. But that doesn't fully explain the subtle magic of being there. Science may illuminate the how, but not always the why. When I think back to that American man's continents-and-sea theory, I can see what he was trying to do with his storytelling. He was working to open up something in me. Maybe that's how meaning works sometimes. Not because it's right or provable, but because it lands somewhere you didn't expect. I don't know if the rocks were meant to be continents, or if the gravel really was meant to be the sea. But for a moment, the garden made room for that idea, and for me. It held me long enough to loosen something, to make me wonder. What matters is that the garden held that possibility – and held me, too, long enough to start asking different kinds of questions. I was curious about who was in the garden and, critically, who was not. Who had access to its benefits? Those who could afford to fly halfway or so around the world. Even the history of the space hinted at a particular kind of exclusion. The temple in which the garden sits began its life as an aristocrat's villa before it became a Zen sanctuary. That lineage – a progression from wealth to religious privilege – made me ask: if these spaces offer such profound benefits (and I'd felt them myself), then isn't it possible that the people who might need them most are the least likely to access them?

*

While I was carrying out my doctoral research, I was struck by the story of Gerrard Winstanley and the Diggers. Winstanley, a radical thinker of the mid-17th century, felt to me like an ancestor of both allotmenteering and guerrilla gardening.[2] Although he would never have described himself in such modern terms, his actions were unmistakably radical: he planted seeds in common land without permission, asserting through action the fundamental human right to cultivate and live from the land.

In 1649, England was reeling from the turbulence of civil war. Amid this instability, Winstanley led a group known as the Diggers, or the True Levellers, onto common land at St George's Hill in Surrey. Their vision was audacious yet simple: to reclaim land for communal use, providing food and sustenance for all people, irrespective of social status. They planted vegetables and grains, intending to create a society based on equality, cooperation and collective stewardship of the earth. For Winstanley and the Diggers, the earth was a 'common treasury', a resource that belonged equally to everyone. Their action was more than symbolic: it directly challenged the prevailing social order, which restricted land ownership – and the food security it provided – to a privileged few.

The Diggers' act of planting without permission was met with fierce opposition from local landowners and authorities, who saw such activities as a direct threat to social hierarchies and what they considered their established property rights. The Diggers faced persecution, violence and, ultimately, expulsion from the lands they sought to cultivate. Yet despite their short-lived experiment, they left an enduring legacy. Their actions, and their example, resonate in contemporary movements that reclaim urban land in order to foster community resilience, food sovereignty and ecological regeneration.

In learning about Winstanley, I felt a sort of kinship – both with him and with those who, centuries later, still see gardening as an act of defiance and hope. The Diggers remind us that beneath our feet, in the soil we share, lies a powerful possibility: a radical vision of coexistence, of humans working alongside nature to thrive together.

For most of human history, land was not something you owned. It was something you belonged to. The woods, fields and pastures were spaces of shared use, shaped by generations who grazed their animals, gathered wood and planted crops. On the British Isles this was called the commons. It wasn't an unblemished utopia. There were rules, hierarchies and disputes. But, crucially, there was access. There was an understanding that land was, in some essential way, collective.

That began to change in Britain with increasing intensity from the 16th century onwards. Through a process known as enclosure, land that had once been held in common was fenced off, deeded and claimed – first by the nobility, then by a rising class of landowners. In the name of efficiency, land was reorganized, rationalized and stripped of its former stewards. Access was curtailed. Rights were revoked. Communities were literally uprooted.[3]

It is easy to talk about enclosure in abstract terms – as an economic shift, a legal sleight of hand – but it was also a profound emotional and physical dislocation. Imagine a child used to running barefoot through open fields, suddenly meeting a fence where there had never been one before. A grandmother who once picked herbs for her tinctures now warned off by posted signs. A farmer staring helplessly as land passed down for generations was divided up as part of someone else's estate. These fences were boundaries of belonging, carved into the land itself.

There's an old protest poem that captures this better than any history book ever could. Its rhythms are simple, its message searing:

> *The law locks up the man or woman*
> *Who steals the goose from off the common*
> *But leaves the greater villain loose*
> *Who steals the common from off the goose.*
>
> *The law demands that we atone*
> *When we take things we do not own*
> *But leaves the lords and ladies fine*
> *Who take things that are yours and mine.*
>
> *The poor and wretched don't escape*
> *If they conspire the law to break;*
> *This must be so but they endure*
> *Those who conspire to make the law.*

The law locks up the man or woman
Who steals the goose from off the common
And geese will still a common lack
Till they go and steal it back.[4]

The poem doesn't need to make an academic argument. It lands where it hurts and where it matters, naming with brutal clarity the injustice baked into enclosure. And it reminds us: taking the commons wasn't just theft, it was authorized theft – enshrined in law, dressed in legitimacy.

The 1773 Inclosure Act and later the General Enclosure Act of 1845 codified and accelerated these changes.[5] And just as the countryside fields were lost so was a way of seeing the land. People who had once lived in relationship with place were transformed into trespassers. The countryside emptied as rural livelihoods collapsed, thus feeding the growth of Britain's cities.[6] But those cities, now swelling with the dispossessed, were not designed for their flourishing.

And so urban green spaces became a new battleground for access and exclusion. Parks were created, yes, but they were often ringed with railings. Walled gardens, once symbols of contemplation and cultivation, became metaphors for who belonged and who didn't. The echoes of rural enclosure lived on in the city – in who had keys to the garden square, who could picnic on the lawn, and who was moved on for sitting too long on a bench.

These patterns persist. Take Freeman's Wood in Lancaster, for example. For years, it was a beloved urban wild space, used freely by locals for walking, biking, foraging. Then, in 2012, fences appeared overnight. Signs warned: *Private Property*. It turned out the land had been surreptitiously acquired by a corporation seeking to sell it for development. A place that had felt, for generations, like part of the neighbourhood's fabric was suddenly off limits. The locals fought back, making maps, gathering memories, pressing legal claims, and they won.[7] But the heart of the conflict was simple: who decides what land is for, and for whom?

Enclosure never really ended. It just moved locations and changed its language. Today, we don't always call it enclosure. Sometimes we call it redevelopment. Or regeneration. Or conservation with restricted access. But the effect is the same: control over land concentrated in fewer hands, and the soundless erosion of public life.

And here's what lingers for me: that the people most in need of sanctuary – those without gardens, front or back, without inherited acres – are often those most excluded from the remaining spaces of green. Whether it's a child staring through iron railings at a private park, or a mother told to move her picnic because she doesn't have the right kind of permit, or a young person warned off for walking a 'suspicious' route through a manicured plaza, the boundary lines remain.

They may no longer be fences and stone walls, but they still say: not for you.

These thoughts aren't new. There have always been people – far more visionary and revolutionary than me – who didn't just reflect on these injustices, but acted to change them.

*

In the grimy quarters of late 19th-century London, greenery was a rare luxury. The Industrial Revolution had transformed the city into a maze of factories and tenements. For the urban financially poor, daily life meant overcrowded rooms, foul air and scarcely a glimpse of nature. Octavia Hill emerged in this era as a social reformer with a radical vision: to alleviate poverty not only through better housing, but through access to green space. Hill had begun her work managing housing for the poor (with support from writer and art critic John Ruskin) in the 1860s, but she quickly realized that bricks and mortar alone were not enough.[8] The children of Marylebone and Southwark, which were considered slums at the time, had nowhere safe to play except the streets, and their parents had no 'open-air sitting rooms' in which to rest from the drudgery of city life. Hill believed that this lack of nature was as debilitating as any material want. She famously wrote, 'The need of quiet, the need of air, the need of exercise: the sight of sky and of things growing seem human needs, common to all.'[9] In one simple sentence, she asserted that contact with nature was not solely a middle-class privilege but a universal right.

Hill's conviction that green space was essential to human well-being was politically charged. At a time when the urban financially poor were often rendered invisible in public life, her insistence that they deserved beauty and breathing space carried a discreet revolutionary force. It was a profound challenge to Victorian society's complacency about inequality. Hill argued that parks and gardens could restore dignity and health to those trapped in London's soot and squalor. In one impassioned appeal, she described how even a few acres of greenery – 'a few acres where the hilltop enables the Londoner to rise above the smoke' – would give city dwellers a chance to 'feel a refreshing breeze that still carries the scent of hawthorn'. Such statements were reflective and compassionate yet rooted in shrewd social observation. Hill understood that poverty was not only an economic condition but a physical and spiritual imprisonment, exacerbated by the grim urban environment. Her remedy blended practical action with an almost poetic awareness of sensory detail: fresh air, birdsong, flowers and sunlight – things sorely lacking in London's slums – that could heal communities.

It was radical enough to suggest that working-class Londoners should enjoy parks; more radical still were those who stepped up to create those parks. In an era when women were expected to remain in the domestic sphere, Octavia Hill and a network of Victorian women reformers boldly moved into the public realm of urban planning. They turned spaces deemed worthless – refuse-choked scraps of urban land, derelict churchyards, the various 'no man's lands' of the industrial city – into green havens. In doing so, these women were waging a peaceful green rebellion against both urban blight and the social constraints on their gender. Hill and her allies operated through philanthropic committees and local campaigns, spheres considered suitably feminine at the time, but they used these platforms to wield genuine power over the design of the city.

One of Hill's comrades-in-arms was Fanny Wilkinson, a trailblazing landscape designer. In 1884, Wilkinson became Britain's first professional female landscape gardener when she was hired by the Metropolitan Public Gardens Association (MPGA).[10] The MPGA was a charity founded two years earlier as part of the wider movement, which Hill had sparked, to green the city's poorest districts. Backed by socially-minded patrons (Hill herself helped establish an earlier group, the Kyrle Society, with similar aims), the MPGA set out to convert disused land into public gardens 'for the benefit of the inhabitants of London'.[11] Wilkinson, then only in her late twenties, took charge of this mission with astonishing vigour. She ultimately designed more than 75 public gardens across London, focusing on areas that, as one writer notes, 'sadly needed the blessing of greenery'.[12] With measuring tape and drawing board in hand, she planned layouts for little pocket parks and large recreation grounds alike. But Wilkinson was no mere armchair designer. Contemporary accounts describe her tramping across muddy lots in her long skirts, supervising work crews and even directing burly navvies (labourers) on how to grade paths and plant trees. It was a conspicuous role for a young woman, and some of the men she hired were reportedly baffled at taking orders from a lady. Yet Wilkinson earned their respect through her expertise and hands-on leadership. The sight of this determined woman transforming derelict corners of the city into oases of flowers and shade was itself a rebuke to Victorian norms.

Crucially, these women reformers understood that public space is political. By physically reshaping the city, they were also redefining who had a right to occupy urban space. Hill often collaborated with female friends and volunteers to organize local tenants, raise funds and lobby officials to support garden projects. The very act of a woman standing up in a parish meeting to argue for a new park or negotiating with church authorities to access an abandoned graveyard for a garden was unconventional. Social reform, philanthropy and early stirrings of feminism were thus intertwined. The historian Gillian Darley notes that Octavia Hill became 'the patron saint' of efforts to improve the lives of the poor through environmental change. Suffragist leaders, too, joined forces in these green campaigns. In one notable case, Millicent Garrett Fawcett – better known for fighting for women's right to vote – led the charge to save an eight-acre green space in London's Vauxhall district from being lost to developers.[13] Fawcett marshalled public support and carried out fundraising, while Fanny Wilkinson drew up plans to turn the threatened land into a park. The success of this effort was celebrated in 1890 when Vauxhall Park opened to the public, complete with a playground for children and sheltered benches for the elderly. At the opening ceremony, the Princess of Wales herself reportedly complimented Wilkinson on her design, as Octavia Hill, Fawcett and other reformers looked on with pride. Such events symbolized a victory both for urban greenery and for women's agency in public life. These Victorian women proved that they could mobilize communities, influence policy and literally change the landscape of the city.

Perhaps the clearest expression of Octavia Hill's vision – and the collaborative spirit of these green rebels – was the creation of the Red Cross Garden in Southwark. In 1887, Hill seized an opportunity to transform a foul industrial ruin into a model urban garden. The site was a wedge of land in one of London's most crowded slums, originally occupied by a paper factory and a row of tumbledown workshops. Hill later recalled her first impression of the place in stark terms: 'It was ... a waste, desolate place. There had been a paper factory on one half ... which had been burnt down. Four or five feet of unburnt paper lay in irregular heaps, blackened by fire, saturated with rain, and smelling most unpleasantly.' The other half of the land was overshadowed by a towering warehouse that blocked light and air from the surrounding tenements. Most city officials would have deemed this plot hopeless, but Hill saw potential where others saw refuse. With her characteristic mix of practicality and idealism, she set to work. Her team of workers laboured for weeks burning the soggy mountains of paper – bonfires blazed day and night – before demolishing the derelict warehouse. 'Our next task was to pull down the warehouse, and let a little sun in on our garden,' Hill wrote, noting with satisfaction that removing the structure provided 'additional light, air and sight of sky to numerous tenants' in the adjacent block.[14] The very elements the slum had been lacking – sunlight, open vistas, fresh breezes – were brought in by clearing the space.

By the following year, in 1888, what emerged in place of the rubble was a small paradise amid the tenements. Red Cross Garden opened with curving lawns, bright flower beds and a playful layout of paths. There was an ornamental pond fed by a fountain, and a bandstand where musicians could entertain on summer evenings. Hill, ever attentive to the needs of children, built a playground, even covering part of it with a wooden arcade (fashioned from salvaged warehouse timbers) so that the children had a dry place to play on rainy days. She also constructed Red Cross Cottages, a row of charming Tudor-style model homes for working families that flanked the garden, each of which had its own small front yard for plants. Next to the cottages, a community hall (Red Cross Hall) was erected to host educational clubs, concerts and even a boys' drill team, reflecting Hill's belief in nourishing bodies, minds and souls. The whole ensemble was an experiment in holistic urban renewal: housing, recreation and nature blended seamlessly. It is little wonder that historians describe Red Cross Garden as Hill's 'pioneer social housing scheme' – a prototype demonstrating how degraded city spaces could be reborn to serve the financially poor.[15]

Newspaper reports from the time marvelled at the transformation. Where there was once a stinking dump, by 1888 there were tidy gravel paths lined with shrubs, and the sound of a fountain mingling with children's laughter. Local residents – many of whom had never before had a green spot to call their own – took pride in maintaining the flower beds and organizing picnics and dances in the hall. Hill was delighted to see the community embrace the garden. She wrote of a 'flower show' held there, noting how Southwark's working families eagerly participated, and describing the burst of colour from the exhibits of home-grown roses and geraniums in an area formerly starved of beauty. Through Red Cross Garden, Octavia Hill proved that even the most marginalized neighbourhood could bloom with a bit of soil, sunlight and social solidarity.

While Red Cross Garden thrived in Southwark, Fanny Wilkinson was busy greening other corners of London in parallel. Her approach was creative yet pragmatic. Wilkinson's projects brought tangible health benefits and new leisure opportunities to London's financially poor. For example, at Meath Gardens in Bethnal Green, Wilkinson transformed an eleven-acre disused cemetery into gardens and playgrounds, employing local labourers in the process. Once regarded as a squalid eyesore, the site was soon celebrated as a welcome and healthful retreat for the area.[16] She also worked on Myatt's Fields in Camberwell – a fourteen-acre tract that provided much-needed open space in a dense neighbourhood. In her largest designs, like Meath Gardens, Myatt's Fields and Vauxhall Park, unemployed men were hired to do the landscaping work. In this way, the act of creating the park also provided jobs and skills – a thoughtful touch of social engineering. But Wilkinson did not neglect the little spaces: a triangle of land beside a railway arch, a quay by the Thames or the forgotten space behind a library – any snippet of land was fair game for improvement. Over the years, she turned such scraps into pocket parks with names like Paragon Gardens and Albion Square, modest green retreats that nonetheless improved daily life for thousands of residents.

By the end of the Victorian era, thanks in no small part to Wilkinson and Hill, the idea that parks were essential urban infrastructure had taken hold. From the 1880s, over a hundred new green spaces were created in London, many in its poorest wards, and other cities in Britain began to follow suit. What had started as an improvised rebellion by a handful of women reformers had grown into a broader urban parks movement.

As Octavia Hill and Fanny Wilkinson were planting seeds of change in central London, their contemporary, Henrietta Barnett, was pushing the green crusade to the outskirts, and into the future.[17] But Barnett's contribution to the movement took a distinct form: she became the driving force behind Hampstead Garden Suburb, an ambitious effort to design an entirely new kind of neighbourhood that would integrate green space and social reform.

Barnett's chance came at the turn of the century, when a large tract of land on London's northern edge was threatened by speculative development. In 1903 she learned that a new railway line (the Hampstead Tube) would open near Hampstead Heath, making the area ripe for building. Barnett feared that the beloved Heath, a wild common where all classes mingled in nature, would be hemmed in and spoiled by 'rows of ugly villas' like those that had overtaken other suburbs.[18] Wasting no time, Barnett organized the Hampstead Heath Extension Council, rallying wealthy and working-class supporters alike to save the land. Their campaign succeeded dramatically: they raised funds to purchase eight acres of fields and woods and in 1906 handed it over to the London County

Council to keep forever as a public open space. This victory – preserving a green commons for Londoners in perpetuity – was very much in the spirit of Octavia Hill's battles to protect urban commons (Hill herself had been involved in saving other pieces of Hampstead Heath in the 1880s). It also foreshadowed Barnett's next step. Having safeguarded one green swathe, Barnett set out to ensure that the inevitable new housing built nearby would embody the ideals of light, air and beauty that she cherished.

The concept of Hampstead Garden Suburb, founded in 1907, grew directly out of the Heath Extension effort. Rather than allow piecemeal suburban sprawl, Barnett envisioned a planned community that balanced development with generous green space. In an article for *The Contemporary Review* in 1905, she outlined a bold social experiment: a neighbourhood where 'people of all classes' could live together in 'a beautiful and healthy place', with nature as an equalizing presence. This was a rejection of the segregated city, where wealthy districts had leafy boulevards and the poor were relegated to smoky slums. In Barnett's suburb, rich and poor would be neighbours by design, sharing the same parks and vistas. To achieve this, she set strict ground rules for planning.[19]

With the help of architects like Raymond Unwin and Edwin Lutyens, Barnett's Garden Suburb came to life over the next decade. Red-brick cottages with lattice windows and climbing roses, allotment gardens for growing vegetables, and curving lanes that respected the natural contours of the land all reflected her influence. By 1911, families of varied income levels had moved into houses that, while different in size, all opened onto the 'green, leafy' environment that Barnett had carefully cultivated. The experiment was widely celebrated. Visitors noted how children from wealthy and poor families played together on the village green and in the woods, something nearly unthinkable in the stratified city. Henrietta Barnett had proved that urban design could be an instrument of social integration. Her work at Hampstead was a capstone to the Victorian era of reform: it carried forward Octavia Hill's dream of healthful spaces for the poor, but on a grand, planned scale. In later years, Barnett's ideas would inspire the Garden City movement and modern public housing designs that incorporate green courtyards and parks. In that sense, she and her peers had sown seeds that would keep germinating long into the 20th century.

The influence of these Victorian green rebels still runs through the cartography of London – the desire to democratize nature stitched into its leafy margins and interstitial green. Yet, through the slow creep of gentrification, some of the very inequalities that these reformers sought to uproot have reappeared, cloaked in new forms. Access to green space – once a rallying cry for justice – is now, in some areas, a postcode lottery. It's a powerful reminder that making space for nature in cities is not a one-off act of benevolence, however visionary, but an ongoing commitment – as cyclical and necessary as the seasons. Renewal requires vigilance. And, perhaps most vitally, access transforms. When green space is genuinely shared, marginalized lives become visible not through pity but through participation – not as passive recipients of aid, but as active guardians of community and place.

*

But the right to space is about more than just access, it's about recognition: not just the freedom to enter a garden, but an invitation to feel at home within it – physically, emotionally, spiritually, politically. Whether it's a Kyoto rock garden, a London allotment or an overgrown wood claimed by a community – these spaces whisper something essential about who we believe matters, and who we imagine cities are really for.

There's a quiet violence in telling people that they don't belong somewhere. Sometimes it's overt: a gate, a fence, a warning sign. Sometimes it's subtler: pricing people out, turning wild places into polished ones, planting hedges that are less about beauty and more about division. But the effect is the same: enclosure doesn't only shrink land, it shrinks possibility.

And yet. There have always been those who resist. Who plant where they're not meant to, who imagine differently, who redraw the maps in our minds. From Winstanley's 'common treasury' to Octavia Hill's belief in the dignity of air and sky, these acts remind us that belonging isn't granted from above – it's cultivated together.

If gardens are where we learn to belong, they're also where we begin to notice who's missing. They show us the edges – who's kept out, who's let in – and ask us whether we're willing to redraw the lines. But it's one thing to contest power when it wears a human face.

What happens when that space is already claimed, not by people but by other species who were there long before we were? What does it mean to belong in a landscape where you're no longer the apex? What kind of relationship is possible when we're asked not just to tend but to live alongside something that has the power to kill us? This is where we will turn next.

Chapter Three

Fear as Coexistence: Power in the Same Space

The car jolts as it swings off the tarmac onto the rocky stones of the Nairobi National Park. Dust puffs up behind us like little clouds. Gary accidentally hits a big bump and I love it. My bum lifts off the seat and for a moment I feel like I am flying. 'Not everyone grows up in a city where lions are your neighbours,' says Gary, as he smiles from behind the steering wheel. I like the way Gary Gunner explains things. He is the coolest friend I have, even though he is really Mummy's friend. He is teaching me how to bird-watch in the park. The binoculars he bought me are ginormous but he says that even though I am nine I'm going to be a serious bird-watcher and so I need to have the right kit. As we continue through the park, the smell of dry grass enters the car and I lean forward looking left and right to see what sort of adventure might arrive before we stop. We are driving to a place in the park where it is safe to get out and walk. Gary says that later in the day is a good time to see a kori bustard or black-headed heron.

Then out of nowhere, a lion appears – calm, confident, like he's on a casual evening stroll. We stop the car, obviously, and I'm watching him do his thing when something catches my eye in the background: glass buildings, Nairobi city centre, rising up behind him like a mirage. After a whole day in the park, it doesn't even feel real. A lion in front of me, skyscrapers behind him. For a second, I just stare. This is wild – literally.

Later that day, I tell my family about the incident with the lion. I'm still thinking about it while Auntie Agnes is doing my hair, getting me ready to go back to school. The lion, the city, and the weird feeling it left behind. It's like I'm stuck between two worlds, the wild and the human-made. But that feeling isn't new. I know it already. I live in it. Right now, I'm on the veranda, knees tucked up, trying not to squirm too much while my aunt oils and parts my hair. I've already tried all the excuses – homework, stomach ache, 'Didn't we just do this?' – but none of them work. Once she starts braiding, that's it.

She doesn't begin with the story right away. She waits. Waits until I get restless and start huffing a bit. Then, without missing a beat, she leans in and says, 'Nyar Rae,* let me tell you something.' That's the moment it changes. Her voice goes soft but serious, like something important is coming. And just like that, I stop moving. Because if there's one thing better than getting out of the eight hours it takes to braid my hair, it's one of her stories.

'A long time ago,' Auntie Agnes begins, 'there was a mother, a mother who had known deep sorrow. She had lost almost everything. She had three children – one after the other, they all left her. The firstborn, stubborn as he was, did not listen to his mother. He ate the bright-coloured seeds of the *mongaluchi* plant [rosary pea]. He fell sick and he died. The second child – eh, that one was taken at birth, didn't even get to cry. And the third, her beloved last-born, just went to sleep one day and never woke up. No one knows what happened. Her husband? He was a good man, but sorrow

---

* My paternal family's nickname for me.

swallowed him whole. One day, he walked away and never came back. And so, this poor mother was left alone. The only thing she had left was her sheep – a beautiful flock. She cared for them well, and they did so well! Fat, strong, healthy – word spread, ah, it spread far and wide. And you know how the world is, Nyar Rae – when you have something good, someone will always be watching and waiting.'

Her fingers never stop moving, parting and twisting my hair.

'That someone was a greedy, wicked hyena. Every night, he snuck in and stole one of her sheep. The poor woman tried everything – ah, she tried! She reinforced her fence with Kei apple thorns, she stood guard with her *fimbo* [staff] – but that hyena was too clever. It was not long until only one sheep remained. And oh, her heart! She feared even that last one would be taken.'

Auntie Agnes let out a small sigh, the way she always did when she was getting to a new beat in the story.

'Then one day, as she sat crying by her gate, a lion passed through the village. A big, strong lion. He saw her tears and asked, "Mama, why do you weep?"

'And so, she told him. She told him everything – her children, her husband, the hyena, the stolen sheep, her fear. And the lion, with a roar that shook the earth, said, "Do not cry any more. I will sort out this hyena."

'That night, the lion hid where the last sheep was sleeping. And when the hyena came creeping through the bushes, greedy as ever, he was not alone. He had brought a jackal along.

'"Where are you sneaking off to, hyena?" the jackal asked, eyes glinting in the moonlight.

'"Ah, my friend," the hyena laughed, "I'm going to take that last sheep. Come, it will be easy!"

'But do you know what they found?'

She leaned in towards my ear and whispered.

'Not a sheep, but the lion himself!

'And so, the jackal ran. But the hyena? Ah, poor fool. He begged, but the lion had no patience for thieves. With one mighty blow – *kwisha!* – that was the end of him.

'And the mother? She never lost another sheep again.'

*

It wasn't until I shared the story with my mother as an adult that I learned that Auntie Agnes was herself a bereaved mother. My mother understood my father to be the lion in the tale (as it was he who brought her to live with us and helped her to start a business). What stayed with me, though, wasn't just the magic of the story – it was the ambivalence and confusion I felt about whether I was supposed to be afraid. Lions were known to be dangerous, weren't they? Yet here was one that showed mercy.

That tension – between reverence and risk, between danger and protection – has always shaped our relationship with lions. Even in cities like Nairobi, where lions are fenced in but never entirely forgotten, fear lingers like an ancestral memory. Folk tales often play with this ambiguity: the lion as monster, saviour, lover, leader. And perhaps that's the point. The presence of a lion forces us to reckon with the limits of our control. It makes us ask whether nature must always be tamed to be safe, or whether living alongside something more powerful than us is part of what makes us human.

We may think of folk tales as 'just stories', but I find that they always have a kernel of truth; and for me it was the compassion the lion displayed. In my child's mind, having grown up in a city that harbours lions, it was not a huge stretch to believe in aunties who were lion whisperers. As it turns out, however, truth is often stranger than fiction, and there are indeed places in the world where lions wander through villages, and in Kenya a lion who cherished what should have been her prey.

*

In the early 2000s a driver working at a tourist lodge in Samburu in the north of Kenya spotted an odd couple in the Samburu National Reserve. It was a baby oryx and a lioness who were living together. Oryx are large, graceful antelopes native to arid and semi-arid regions of Africa and the Arabian peninsula. They are known for their striking appearance, featuring long, straight or slightly curved horns that can grow up to several feet in length, and distinctive markings. They are also usually preyed on by lions. This time, however, the lioness was looking after the baby oryx, and they were seen licking each other as if they were kin rather than foe. The lioness went on to be named Kamunyak, which means 'the miracle or blessed one' in the Samburu language. Although Kamunyak's adopted oryx was eventually killed by a male lion, she still went on to adopt another four. This did not surprise local Samburu peoples, who spoke of a three-year-old child who had previously been found happily playing with lion cubs.

As much as the child in me wishes it were so, it would be naïve to imagine a world where humans and wild animals live together in perfect harmony. That version of nature – peaceful, passive, benign – exists mostly in picture books and fantasies. In reality, coexistence is far more complicated. It involves risk, tension, and a negotiation of boundaries that are constantly shifting. And yet, I've never been able to shake the feeling that nature, despite everything, leans towards balance. That somewhere between conflict and peace, fear and reverence, there are moments – brief and startling – that offer glimpses of something gentler. Something that feels like magic.

Lions, in particular, carry that tension with an almost mythic weight. They are apex predators, animals to be feared, and rightly so. But they've also long been symbols of something else: nobility, guardianship, even grace. In the Biblical tradition, the story of Daniel in the lions' den presents this contradiction in full force. Daniel is thrown into a pit of lions for refusing to obey the king, and yet he emerges unharmed. The lions, instead of attacking, remain still – their mouths closed, their power withheld. Whether read as divine intervention or metaphor, the story offers a provocative question: what does it mean to sit beside something powerful and not be destroyed? Can the wild be near us, even protect us, without being tamed?

These aren't just theological musings. They strike at the heart of how we think about fear – and what it means to live alongside animals that could kill us, but don't. It's that uncertainty – the razor-thin line between danger and intimacy – that makes the lion such a potent figure in our stories, and in our real lives. And nowhere have I seen that contradiction more vividly than in the story of Kamunyak.

Daniel was not the only human to be spared by lions. Among the tales attributed to Aesop – who may or may not have been an African slave living some time between the 1st and 6th century BCE – is the enduring story of an enslaved man who earned his freedom through an extraordinary bond with a lion. The story goes that Androcles, a slave suffering under the cruelty of his master, decided to escape and flee into the forest. Exhausted, he stopped to rest, but soon found himself face to face with a lion roaring fiercely. Terrified, Androcles believed his end had come. Yet, as he watched, he realized that the lion's cries were not from hunger but from pain – a thorn was lodged deep in its paw. Summoning his courage, Androcles approached the lion, gently removed the thorn, and cleaned the wound. To his surprise, the lion responded with calm gratitude rather than aggression.

Not long after, Androcles was captured by his master and thrown into prison. As punishment for his escape, he was sentenced to face a lion in the grand arena, a spectacle to entertain the emperor and the crowd. When the fateful moment arrived, Androcles was led into the arena to meet his doom. But instead of attacking, the lion recognized Androcles as the one who had helped him in the forest. Instead of roaring, it approached him affectionately, much to the astonishment of the crowd.

The emperor, moved by this display of loyalty and friendship, declared Androcles and the lion free. The two were released together, their bond serving as a timeless reminder of the power of compassion to bridge even the widest divides.

In another Greek legend, the tale of the Nemean lion again offers a profound example of how myths and stories shape our understanding of the natural world and the forces that govern it. This legendary beast, said to be invulnerable to mortal weapons, became a symbol of strength and perseverance, reflecting humanity's enduring fascination with the mysteries of the animal kingdom and its place in our shared narratives.

Once upon a time, the people living near the valley of Nemea in southern Greece were terrorized by a fearsome lion that roamed the nearby hills, preying on anyone who crossed its path. It was a powerful creature with skin that no weapon could penetrate. And it was precisely this skin that King Eurystheus (king of Tiryns) demanded that Herakles (better known as Hercules) should bring to him. It was the first of the twelve impossible labours that Hercules had to perform as punishment for having murdered his own wife and children.

Before setting out on his seemingly impossible task, Hercules stopped in the small town of Cleonae, where he was welcomed by a poor workman named Molorchus. Eager to support Hercules, Molorchus offered to sacrifice an animal in hopes of a safe hunt. However, Hercules asked him to wait for 30 days. If he returned with the lion's skin, they would offer the sacrifice to Zeus, king of the gods. But if Hercules perished in the attempt, Molorchus agreed to make the sacrifice to Hercules as a hero instead.

When Hercules reached Nemea and began tracking the lion, he quickly realized that his arrows were useless against its impenetrable fur. Undeterred, Hercules picked up his club and tracked the beast to a cave with two entrances, where he blocked one before approaching the lion through the other. In a fierce battle, Hercules grasped the lion in his powerful arms and, despite its vicious claws, choked it to death with sheer strength.

On the 30th day, Hercules returned to Cleonae with the lion's carcass. Molorchus, relieved to see him alive, joined the hero in sacrificing to Zeus. Hercules is often depicted in ancient Greek art wearing a lion skin, a symbol of his victory.[1] While some writers believe this iconic hide came from the Nemean lion, others suggest that it was from a different animal slain by Hercules in his youth. The playwright Euripides claimed that the skin came from the sacred grove of Zeus in Nemea.[2]

It would be easy to regard the lion in Greek mythology as nothing more than a fantastical beast, a creature of heroic tales, far removed from the real world. But mythology often grows from generations of remembering, and recent archaeological finds suggest that these stories may be grounded in a reality where lions were encountered in day-to-day life. Excavations around the ancient city of Tiryns have uncovered skeletons believed to belong to *Panthera leo*, prompting scholars to reconsider how closely humans once lived alongside these animals.[3] Some argue that they were exotic imports, trophies of power kept by elites, but that theory becomes harder to support when we see further lion remains turning up across south-eastern Europe, including in modern-day Bulgaria. These weren't isolated spectacles. They were animals with a foothold in the landscape.

And so we're left with a striking possibility: that lions once moved through the same spaces as ancient Greek communities, not as zoo animals or passing curiosities, but as part of the ecological and cultural fabric. What would it have meant to fetch water, herd goats or sleep at night knowing a lion might be watching from the hills? The stories that grew from those encounters – of battles, bravery and monstrous strength – tell us as much about human fear as they do about lions themselves. But the fact that people didn't simply flee or wipe them out speaks to something else too: a begrudging familiarity, maybe even respect. As with so many lion stories, the boundary between terror and reverence was thin. And that, perhaps, is the true heart of coexistence.

The archaeological record provides tantalizing clues suggesting that ancient European cities, like Nairobi today, might have shared their spaces with lions. Recent discoveries in Bulgaria, led by Nadezhda Karastoyanova from the National Museum of Natural History, have uncovered evidence of lion hunts dating back over five millennia. These indigenous lions were not only hunted but also brought back to settlements, possibly as food or as symbols of power and prestige. This underscores the close relationship between humans and these predators in what could be described as urban or semi-urban settings. Lions, it seems, were not seen solely as creatures of the wilderness but as beings entwined with human life in ways that were both practical and symbolic.[4]

Further evidence comes from Durankulak on the Black Sea coast, where lion remains have been uncovered near prehistoric settlements. Even more striking is the discovery of a prehistoric human skeleton at Kozareva Mound, bearing injuries probably inflicted by a lion. Remarkably, the individual survived and healed from these wounds, a confirmation of the resilience of humans living alongside formidable predators, and a suggestion that lions were not distant threats but were actively integrated into the human world, cohabiting with early communities and becoming part of their urban ecosystems.

These archaeological discoveries not only validate the historical presence of lions in Europe but also invite reflection on how they shaped human stories and human strategies. In ancient Greece, lions were real, proximate threats, and their presence demanded a response. In the myth of Hercules and the Nemean lion, victory didn't come from brute strength alone, but through cunning and adaptation. Hercules had to study the lion's movements, learn its strengths and find a way to outsmart it. The lesson buried in that myth is still relevant: to live alongside lions – or any powerful force of nature – humans must learn how to work towards balance and not simply move to eradication because of fear.

Greek mythology, art and literature reveal more than symbolic associations. They reflect a time when lions were part of the shared environment, part of the mental and physical geography of daily life. And in that world, fear was not an overreaction, it was a requirement. To fear the lion was to survive it. But fear did not mean exclusion or extermination. Instead, it shaped a form of coexistence based on respect, attention and a certain kind of humility. The lion demanded distance, yes, but also recognition – of its power, its role, its presence. And in that uneasy balance between awe and anxiety, a pattern emerges that echoes through time: wherever people have lived with lions, they have done so not by eliminating fear, but by learning how to live with it.

*

African cultures share a rich tapestry of narratives that prominently feature lions, often symbolizing courage, survival and the journey into adulthood. Among Maasai peoples, lions have long held a deeply significant role in one of their most celebrated rites of passage. For young men aspiring to the revered status of moran, or warrior, the act of killing a lion was both a test of bravery and a demonstration of their intimate understanding of these animals. This rite of passage was never undertaken lightly; it was rooted in respect for the lion as a formidable and symbolic creature of the savannah.

It was a dangerous, solemn undertaking, and not every young man attempted it, nor did every attempt succeed. Yet the tradition existed alongside healthy lion populations, as it was practised with a recognition of the delicate equilibrium of humans and wildlife. As a child, I met men who had passed this test, proudly showing off a lion's tooth as a symbol of their accomplishment and their connection to this cultural heritage.

Maasai peoples have lived in and around what is now Nairobi for a long time, and their deep ties to the land are reflected even in the city's name. Derived from the Maasai word *Nyrobi*, meaning 'place of cold waters', it references the swamps and streams where their cattle once grazed. Living in harmony with lions and other wildlife on these grasslands, their story is one of balance and shared space, an enduring lesson for modern conservation efforts. While Nairobi's urban sprawl has pushed much of this history to the margins, Maasai peoples' legacy remains embedded in the landscape and the stories we continue to tell about it.

In recent years, Maasai peoples have adapted to changing realities. With lion populations dwindling, they no longer kill the animals as part of initiation rites. As a moran friend once told me: 'We may no longer use them to become morans, but we still know where they are – we still live with them, just differently.' This shift reflects a profound understanding of ways of being that consider non-human neighbours, one based on respect and observation.

A striking example of this is Richard Turere, a young Maasai man who grew up among his pastoral community in Kitengela, which borders the Nairobi National Park. There is a level on which I can relate to Richard, having also grown up in an area that bordered the park, so I know to an extent what it is like to live with the tension of human–wildlife coexistence. The difference is that my family, being mainly settled, no longer practises all of our pastoralist heritage, whereas in Richard's case his family's livestock are central to their lives and livelihoods.

To understand the importance of cattle within Maasai culture, one must look past their financial value – a hundred head of cattle, for example, is not an insignificant amount of money – as Maasai peoples also have a deep spiritual connection to these animals. They believe that when the earth was young, Engai (or Enkai, their god) lived in unity with the earth and sky and that all the cattle in the world belonged to Engai. One day, there was a powerful volcanic eruption that separated the earth from the sky, taking Engai, and the cattle, to the heavens, and leaving the people alone on earth. Cattle, however, still needed to eat grass so Engai sent all the cattle down to earth via the roots of the sacred fig tree to be looked after by Maasai peoples, a duty they have faithfully upheld ever since.

For Maasai peoples, cattle represent life itself, providing milk, which is central to their diet, hides, for clothing and bedding, and dung for building homes. Even the act of slaughtering a cow is accompanied by prayers and rituals, acknowledging the sacredness of the animal and expressing gratitude to Engai. Wealth is not measured in currency but in the number of cattle one owns, and the exchange of cattle solidifies marriages, friendships and peace treaties.

Spiritually, cattle are seen as intermediaries between Maasai peoples and Engai. The colour of a cow, the way it behaves, or even the circumstances of its birth can carry spiritual significance, interpreted by elders and diviners as signs from Engai. Ceremonies, from birth rites to warrior initiations, are often accompanied by the blessing of cattle, their milk, or even their blood, reinforcing the belief that these animals are not merely possessions but sacred companions in the journey through life.

Because of this connection, Maasai peoples historically viewed cattle raiding not as theft but as a form of reclaiming what they believe Engai had originally intended for them. Although modern laws and borders have altered these practices, the spiritual and cultural weight of cattle remains central to Maasai identity, guiding their values of communal living, respect for nature and the preservation of their way of life against external pressures.

From an early age, Richard understood the weight of protecting his family's livestock, both from a financial and a spiritual perspective. Cattle taken by lions were not only a threat to his family's economic stability but also fuelled a cycle of conflict between local communities and the endangered lions. In his early teens, Richard faced a pivotal challenge when his family lost several cattle to lion attacks. Traditional methods of protecting livestock, such as bonfires and scarecrows, had failed repeatedly. His community was torn, they understood the wider value of lions to Kenya's ecosystem and the tourist economy, but they also had their sacred duty and livelihoods to maintain. Frustrated by this existentialist crisis, yet determined, Richard began to use the skills that have seen Maasai peoples thrive on the savannah for hundreds of years. He closely observed lions' behaviour and noted that they avoided areas with moving lights. This insight sparked an idea that would soon transform his community and garner international recognition.

Armed with old car parts, a motorcycle indicator box and a solar panel, Richard constructed a system of flashing lights that mimicked human movement. Dubbed 'Lion Lights', the set-up effectively deterred lions from attacking livestock, as the predators' believed that people were patrolling the area. The innovation was simple yet revolutionary, reducing human–wildlife conflicts significantly and saving the lives of both livestock and lions. Richard's success spread beyond his village, earning him a platform on the global stage. At just 13 years old, he was invited to share his story at TED Talks, where his compelling speech resonated with audiences and showcased the profound impact of youth-driven innovation. His initiative has since been replicated across various parts of Africa, helping communities coexist more peacefully with wildlife.[5] Richard's Lion Lights are grounded in traditional African natural history practices, where young people learn how to observe and make sense of their environments.

What moves me most about his invention is that it doesn't deny the fear lions evoke – it honours it. The lights don't try to overpower the lions, or banish them, but instead create space: a signal, a boundary, a kind of truce.

In this discreet exchange between light and shadow, danger and safety, I see a different kind of harmony, one that recognizes fear and develops the courage to live with it.

*

Today, we are living in what scientists call the Anthropocene, a geological epoch defined by human dominance over the planet. Through our industries, cities and emissions, and through our endless expansion, we've become the single most powerful force shaping the earth's systems. The balance has tipped. Where we once feared nature, it is now understandable that nature probably fears us, or at least works to flee from or counter our impact. The Anthropocene tells the story of how we moved from living alongside nature to remaking it in our image. And even as we rush to build solutions – green tech, climate accords, protected zones – we often fail to reckon with the truth: that we were part of the problem to begin with.

This, to me, is the paradox of Nairobi National Park: a place that, on paper, stands for protection but whose very existence is entangled in histories of displacement, enclosure and human control. It holds beauty, yes, and wonder, but also silence, compromise and the uneasy knowledge that what we now work to conserve we once took away.

The 1904 and 1911 Maasai Agreements forced Maasai peoples from their ancestral lands into 'native reserves', a hallmark of Kenya's colonial land policies. Negotiated under duress and against a backdrop of devastation caused by rinderpest and other epidemics, these agreements disrupted Maasai pastoral practices and worked to sever their deep ties to the land.[6]

In addition to livestock disease, Maasai peoples were afflicted by smallpox and other epidemics that further weakened the population and made them vulnerable to colonial encroachment. These health crises were compounded by the brutality with which land was appropriated.

The colonial administration, under the guise of treaties and legal agreements, enforced the relocation with military power when necessary, demonstrating the violent nature of their appropriation efforts. Maasai peoples were moved to less fertile and more arid regions, which severely limited their traditional practices of pastoralism and disrupted their socio-economic structures.[7] This dislocation not only affected their access to land and resources but also disrupted their ability to manage and coexist with wildlife. For Maasai peoples, whose traditional practices had supported a balanced ecosystem for centuries, these lands held cultural and economic significance, as well as ecological value.

The land taken from Maasai peoples was among the most fertile in the region and was quickly claimed by European settlers, who sought to develop ranches and large-scale agricultural enterprises. These settlers viewed Maasai peoples as obstacles to progress, dismissing both their rights and their traditional knowledge of land management. The marginalization that ensued was not just about land loss. It also meant an erasure of identity and autonomy as colonial policies were explicitly designed to subordinate local communities and prioritize settler prosperity. The creation of protected areas, such as Nairobi National Park, occurred within this context of disenfranchisement and land expropriation. Its establishment was a direct response to the escalating tensions between European settlers, local communities and wildlife, set against the backdrop of British colonial policies that prioritized wildlife conservation and resource exploitation over the rights and well-being of the African peoples who had lived in the area long before the settlers' disruptive arrival.

The rapid appropriation of Maasai land displaced communities, fuelled resentment and set the stage for mounting human–wildlife clashes as settlers expanded their farms and ranches. By 1910, Nairobi's population had reached 14,000, and encounters with wildlife were common. Colonial residents often armed themselves at night against lions, while complaining of giraffes and zebras trampling their gardens. This uneasy coexistence foreshadowed tensions that still shape the city today.

A century later, in 2016, six lions wandered out of Nairobi National Park and into the neighbouring areas of Kibera and Lang'ata. In Kibera, a densely populated area with limited resources, the lions passed through without incident.[8] In Lang'ata – a wealthier suburb and home to many descendants of British settlers – Kenya Wildlife Service attempted to dart and capture one of the lions but fatally wounded it instead.[9] The contrast highlights the socio-economic dimensions of human–wildlife conflict in Nairobi, where financial resources and prevailing attitudes towards wildlife often influence how interventions play out.

The fascination with big-game hunting in colonial Kenya is a crucial part of this history. The country's abundant wildlife attracted prominent hunters from around the world, including figures such as former US president Theodore Roosevelt and British prime minister Winston Churchill.[10] Both statesmen embarked on extensive hunting expeditions in Kenya, with Churchill visiting in 1907 and Roosevelt following in 1909. Their visits were partly facilitated by E Wellesley Ashe, an Irish-born ranch owner who wrote to Roosevelt, encouraging him to come to Kenya to hunt 'game'. 'Dear Sir,' he wrote to Roosevelt,

> I have heard that it is your intention to visit this country for Big Game hunting and I beg to state that I have a 6000 acre farm in the very heart of the best game country here which I would like to place at your disposal. Most of the African game is to be found on my land including Lion, Leopard, Hippo, Rhino, Giraffe, Zebra, Buffaloe and the various species of Buck. I am bounded by two large rivers and using my Shamba as headquarters you can get every East African animal, with the exception of Elephant, in less than a twelve hours journey.

I have also an option on my two neighbours land comprising eight thousand acres more. Mr. Churchill's party lately killed a lion in this locality.

Should you really contemplate this visit I shall be very pleased to make any arrangement you wish for your reception.[11]

During Roosevelt's hunting expedition, African individuals were regarded as little more than resources to achieve personal ambitions. Roosevelt's hunting party, which also collected specimens for the Smithsonian Institution, amassed approximately 11,400 animals, including 17 lions, 11 elephants, and 20 rhinos.[12] The expedition relied on over 600 African porters, many of whom faced hazardous conditions, particularly those tasked with carrying rifles and engaging directly with dangerous wildlife. Some porters were mauled, and a few even lost their lives, highlighting the risks they were forced to confront in fulfilling the demands of an expedition that prioritized Western scientific and recreational goals over local safety and well-being.[13]

In stark contrast to the extractive logic of colonial expeditions like Roosevelt's, pre-colonial African communities lived in closer harmony with wildlife, guided by cultural and spiritual frameworks that recognized animals as part of a shared world. Animals were hunted, yes, but usually within seasonal and spiritual boundaries that helped maintain ecological balance. Fear, in that context, wasn't something to be eliminated but rather understood – a natural check on how humans moved through the landscape. But with the arrival of big-game hunters and the rise of trophy expeditions, this balance was violently upended. Wildlife was no longer part of a relationship – it became a resource, a commodity, a backdrop to imperial adventure.

This shift towards exploitation didn't just threaten animal populations, it introduced a new kind of conflict. Animals that had once inspired reverence or healthy caution were now pursued to the brink of extinction. In turn, fear was replaced with control, coexistence with domination. The colonial legacy redefined how people related to nature, not as something to live with, but as something to own or subdue.

Nairobi National Park encapsulates this contradiction. Though established as a sanctuary for wildlife, it emerged from a history shaped by dispossession and extraction. It represents both an attempt to preserve what colonial exploitation nearly destroyed, and a reminder that conservation, too, carries the weight of its origins.[14]

Amid the devastation, some voices within the colonial machine began to have their doubts. Mervyn Cowie, Nairobi National Park's first warden, embodied this shift. As a young man, Cowie had hunted in the same savannahs he would later fight to protect. His growing horror at the scale of wildlife loss catalysed his mission to preserve Kenya's natural heritage. Under his stewardship, Nairobi National Park was established in 1946 – not without controversy, as its creation displaced Maasai peoples and other local communities whose ancestral ties to the land were ignored. The park's birth, though groundbreaking for conservation, carried the imprint of colonial arrogance, sidelining the very people who had nurtured this land for generations.[15]

But the conflict did not end with colonialism. Today, the park stands in uneasy relation to the city that surrounds it – a city growing fast, stretching wide, and often forgetting the fragile, breathing space it envelops. Nairobi's expansion has not only pressed up against the park's edges, it has begun to seep into them. Wildlife corridors that once allowed animals to migrate freely are now choked by highways, railways and concrete. Elephants and lions navigate fences. Behaviour shifts in ways we barely understand.

The pressures are both spatial and political. For the park to endure, it needs more than environmental protections on paper. It needs an atmosphere free from corruption, where land isn't quietly sold off under the guise of development, and where decisions are made with both humans and animals in mind. This demands the kind of policy that includes design: an ethic that imagines futures where cities don't swallow everything in their paths.

And then there's the deeper, quieter loss – one harder to measure but just as urgent. The oral histories and traditional knowledge that once shaped how people lived with animals are no longer passed down in the same way. Generations are growing up in a city where lions are roadside billboards or TikTok clips, not a living presence. Few are taught that this urban space sits on layers of story and struggle, that Nairobi is not just a city, but a contested ecosystem. Without that understanding, it becomes easier to believe that wildlife can simply be moved, boxed up, relocated. But the park is not a zoo. It is what remains of a relationship, and our coexistence with Nairobi National Park is contingent on whether or not we remember what it truly means. And there are those who do remember. Today, Maasai communities continue to graze cattle even near the edges of Nairobi National Park – a quiet but powerful assertion of belonging. Their presence challenges the idea that urban conservation and traditional land use must be at odds. It is a reminder that the land has always been lived in, tended to, understood. Richard Turere's Lion Lights, born from this very landscape, show what happens when traditional ecological knowledge and modern technology meet with care. These stories remind us that coexistence is still possible because people continue to live with conflict – creatively, respectfully and with a deep memory of what came before.

*

Globally, the lessons of coexistence resonate far beyond Nairobi. Around India's Gir forest, Maldhari pastoralists live alongside Asiatic lions in a relationship steeped in cultural pride and respect. These are the people to whom I was referring earlier, when I talked of lions wandering through villages. Asiatic lions (*Panthera leo persica*) are a subspecies of the African lion and the very same species that would have lived in ancient European cities. Their territory once covered all of northern Africa, Southwest Asia and Greece. Now reduced to several hundred in number, their existence is intimately linked to their acceptance by the Maldhari peoples with whom they share the Gir forest, an area slightly smaller than Greater London. Despite occasional livestock losses, Maldhari peoples see the lions as guardians of their forest, embodying a harmonious balance that conservationists worldwide strive to replicate. A recent study found that over 76 per cent of people in the Gir landscape expressed positive attitudes towards lions, even as the big cats roam beyond the boundaries of protected areas into human settlements.[16] Their way of life shows that true tolerance is born from deeply rooted cultural values that are woven into daily practices and traditions rather than policies dictated from above.

Lions themselves have shown remarkable adaptability in urbanized and human-dominated landscapes. They have been known to shift their hunting patterns to night-time in order to avoid human activity.[17] In Mozambique, lions' ranges extend across 78 per cent of the country, intersecting with villages and towns.[18] These animals, often considered as threats, can also reveal themselves to be partners in a shared existence, adapting, as must we, to a world where boundaries between the wild and the urban blur.

I think equating coexistence with harmony can be misleading. Harmony suggests a kind of ease, a gentle fit – but living alongside lions, or any powerful animal, demands more of us than comfort. Coexistence requires restraint, attention and imagination – the ability to see another species not as a danger or inconvenience, but as a presence that shapes how we move, build and belong. It asks us to let go of control without giving up care.

If there is wisdom in the stories we tell about lions, from the myth of Hercules to the true tale of Kamunyak, it's that fear can be part of respect; that we do not have to be unafraid to live well with wildness. But we do have to be willing.

But what if fear isn't the only thing keeping us apart? What if it's something more familiar – power, protection, even preference? Sandra McPherson's poem 'Lions' offers a different kind of mirror. Instead of being a moral tale it is a meditation on what lions are and what we project onto them. It doesn't flinch. And maybe that's why I return to it:

> *Lions eat communally but completely lack table manners.*
> *Indeed, lions give the impression that their evolution*
>
> *Toward a social existence is incomplete – that cooperation*
> *In achieving a task does not yet include*
>
> *The equal division of the spoils.*
> *… Their courage comes*
> *From being built, like an automobile,*
>
> *For power. A visible lion is usually a safe lion,*
> *But one should never feel safe*
>
> *Because almost always there is something one can't see.*
> *Given protection and power*
>
> *A lion does not need to be clever.*
> *…*
> *If we made of ourselves parks and placed the lion*

In the constituent he most resembled
He would be in our blood.[19]

The lion doesn't need us to like it. But it does require us to know ourselves. Coexistence doesn't mean seeing lions as noble or evil, just real creatures to have a relationship with; a recognition that they live, breathe, hunger, guard and rest on the same earth as us – sometimes, just on the other side of a road. The future of places like Nairobi National Park will not be secured by fences alone, but by whether we are brave enough to share space, even when it feels uncomfortable. Even when it costs us something.

What I've come to understand is that coexistence is a kind of commitment. It means making room – physically, politically, emotionally – for lives that are not ours. For the lions who guard stories that we are still learning how to read. For the animals that force us to reckon with what kind of humans we want to be. The question isn't whether nature belongs in the city, but whether the city can remember that it was once, and still is, part of nature.

And yet, sharing space isn't always about grand gestures or apex predators. Sometimes, it's about noticing who's already thriving beside us. Coexistence is tested in moments of high drama and shaped in the calmer ones too. The flutter outside your window, the shadow on a wire, the sudden burst of song in the middle of a city morning. If lions stretch our imagination of what's possible when we live with fear, birds remind us what's possible when we live with attention. Both ask us to change – but only one waits on the windowsill.

Chapter Four

Hawks, Ibises and Other Sky Neighbours

I'm sitting on my parents' veranda – the same one where Auntie Agnes used to braid my hair. The only thing that's changed, really, is me. I'm in my thirties now, back in Kenya, trying to give my daughter a taste of the astonishing childhood I was lucky enough to have.

She's latched on to my breast, all warm and drowsy, and I'm lost in the rhythm of feeding when a flash of movement catches my eye: one of my mother's guinea fowl (*Numida meleagrisis*) wandering through the garden, her chicks trailing behind her like soft little commas. Guinea fowl aren't considered the most glamorous of birds – speckled feathers, awkward gait, prone to panic at the smallest thing – but there's something about this one that I can't stop watching. Maybe it's the timing. She and I have become mothers in the same season. My daughter was born during *masika*, the long rains. It's a season of green abundance, of births and beginnings, and something about that forms a link between us.

I find myself feeling oddly protective of this guinea fowl, as if our shared motherhood gives us some kind of unspoken bond. She clucks gently to her chicks, guiding them through the garden, and I imagine us both navigating our little broods through a world that can be so unpredictable. Of course, I like to think that I've got a bit more brainpower than she does. Guinea fowl are famous for their confusion. I've seen them run head first into fences they could easily fly over, calling out in alarm instead of just solving the problem. It's almost comical – unless you're the one trying to get them out of trouble.

Just as I'm about to settle into this gentle moment, a shadow slices across the ground. I look up. It's a black kite (*Milvus migrans*) – what we call a hawk in Nairobi. I've known them since childhood, but I still feel that jolt every time one appears. They float like spirits above the city, quiet and sure. This one is circling, its eye locked on the guinea fowl chicks.

My whole body tenses. I want to leap up, to shoo the hawk away, to gather the little ones and carry them to safety. But I can't – my daughter is still nursing, and anyway, I'm frozen. The kite drops without warning. One strike. You are taken

> *into the talons of Kite*
> *so swift*
> *you, your mother and I*
> *are caught on a breath*
> *and do not cry out.*[1]

Too heavy to carry off, the hawk begins to eat right there, in the middle of the garden. I watch in silence, stunned by the speed, the precision, the brutal necessity of it all. In minutes, the hawk is done. It lifts into the air and disappears.

When I finally get up, after burping my daughter and laying her gently in her cot, I walk out to where it happened. There's nothing left but a few scattered feathers – black and white like the guinea fowl herself – fluttering in the breeze like a memory.

What surprises me most isn't the violence. It's how natural it all is. The hawk didn't kill out of malice. It didn't waste a single morsel. Everything was used. There's something unsettling about that kind of efficiency – but also something honest. In this garden, life ends and begins again, often in the same breath. As a mother, I know the fragility of new life. But standing in that patch of grass, I feel something else too. This wasn't senseless. It was survival. It was the world, doing what it's always done, even while we cradle our children and hope it will pause, just for us.

As I return to the veranda, I'm struck by the intricate connections that played out in front of me, the guinea fowl and the hawk, two lives briefly intersecting in this patch of garden. One is looking for safety, the other for sustenance. And I realize: connection begins with attention. I wouldn't have noticed any of this if I hadn't been watching closely and worked to really see what's moving and what's at stake.

※

Hawks or black kites are raptors of the family Accipitridae and are one of the most abundant raptor species in the world, being found across Africa, Asia and Europe. This is due to their high adaptability, and to the connections they've forged with humans by learning how to live alongside us in cities. Black kites are medium-sized birds with dark brown plumage that can appear almost black from a distance. They have a wingspan of up to 140 centimetres (4½ feet) and a distinctive forked tail that aids their agile flight. Although they are largely carnivorous, eating small mammals, insects and carrion, they are resourceful, which means they will happily frequent rubbish dumps and other food-rich environments.[2]

At gatherings in Kenya where *nyama choma* (roast meat) is a staple, someone invariably tosses a bone into the air, only for an open-mouthed hawk to dive acrobatically and snatch it mid-flight before soaring away to enjoy it. They do not even have to land to do so: black kites are capable of feeding mid-air, using their beaks to pluck the food from their talons. Black kites are scavengers, much like rats or pigeons, who have learned to thrive on humanity's carelessness with food. In Nairobi, they've moved in en masse. Ironically, if you're a bird-watcher, you're more likely to spot a black kite in urban areas than in the countryside.

Watching black kites snatch food in Nairobi is more than just an entertaining spectacle. It hints at a deeper relationship between humans, birds and fire that has been forged over millennia. While in Kenya smoke and flames from nyama choma draw in black kites who are eager to scavenge, in other cultures birds play a more profound role, becoming active participants in the very creation and harnessing of fire itself. One of my favourite examples comes from Aboriginal Dreamtime stories in Australia, where a bird known as the Firehawk is celebrated for its agility or resourcefulness as well as for gifting humanity with fire.[3] The story of the Firehawk, rooted in the rich tradition of Aboriginal Dreamtime, tells of the origins of fire and its transformative gift to humanity. For the Aboriginal peoples of Australia and Torres Strait Islanders, the Dreamtime represents the creation period – a time when ancestral beings, both human and animal, shaped the land, its features and its laws. These stories are deeply intertwined with the landscape, offering a vision of meaning, identity and belonging.

In the Kimberleys, a region of western Australia, one such story recounts how the crocodile, a strong and sullen figure, tried to keep the secret of fire-making to himself. He hoarded the fire sticks, which were the key to creating fire, keeping them hidden under his armpit. Without fire, people were forced to eat raw food, living off lily bulbs, seeds and fish from the rivers and billabongs. Many had tried to take the fire sticks from the crocodile, but his vigilance and ferocity made it impossible.

Then along came the Firehawk, and, observing the people's plight, he decided to act. In a swift and daring dive, he snatched the fire sticks from the crocodile and soared through the trees, too fast for it to catch him. He delivered the fire sticks to an old man called Naga Cork, who showed the people how to use them to create fire. For the first time, the people could cook their food, transforming their lives and marking a critical step in their development. The crocodile retreated to the water, bitter and defeated, where it has lived ever since.[4]

This tale is about more than fire: it's about culture, civilization and our deep interconnectedness with the animal world. Fire brought warmth, safety and the transformative ability to cook food, all of which set humans apart from other animals. Yet this story reveals something crucial: becoming fully human depended entirely on the intervention of animals and the natural world. In other words, the very foundations of human civilization are inextricably linked to, and reliant upon, nature. This is where the role of Naga Cork, the wise elder, becomes pivotal – sharing the knowledge and guidance needed to harness the fire's power. What's fascinating is the Firehawk's instinctive understanding of where to seek this wisdom, emphasizing an innate recognition of the value of collaboration. In modern-day Delhi, this entangled relationship between humans and scavenging birds continues. The Oscar-nominated documentary *All That Breathes* (2022) follows two brothers who run a makeshift sanctuary for black kites injured in the city's polluted skies. Their quiet, determined care reflects a deep recognition: that even so-called nuisance species can become companions, healers or witnesses to our own environmental choices. Much like the Firehawk who brings fire to humans in Dreamtime lore, the kites in Delhi move between danger and salvation and in that way embody both the precarity and possibility of coexistence.

We often overlook how deeply dependent we are, and have always been, on nature, not just in terms of our civilization's origins, but its continued maintenance. Just as the Firehawk story highlights nature's foundational role in human development, urban ecosystems reveal how wildlife actively sustains our daily lives, quietly managing our shortcomings. One compelling example is the essential service that animals such as marabou storks and vultures provide in waste management. These birds perform a vital yet often unnoticed role, tirelessly tidying up after us, consuming waste and carrion, and thereby safeguarding our cities from disease and decay. Their presence underscores both nature's resilience in adapting to human-influenced environments and our critical, if frequently ignored, reliance on their ecological contributions.

In Nairobi, the marabou stork (*Leptoptilos crumenifer*) has become a particularly conspicuous resident, often seen perched atop trees such as the jacaranda and acacia. These large birds have a complex relationship with the city's human inhabitants. While some residents lament the mess left by their droppings – marking clothes lines and city corners in white – others see the marabou storks as neighbours: striking, if unconventional, members of the urban ecosystem, as much a part of the city as any human inhabitant. When trees in which the storks had formed their colonies were cut down, some Nairobians decried it using the same emotive language about the city council's actions as they would for humans: the birds, they said, should not be 'evicted'.

The Kenya Wildlife Service has worked hard to educate Nairobi's citizens about these birds. The marabou storks are unlikely to leave the city any time soon, having become more prominent because of the destruction of their tree homes elsewhere and the city's abundance of food, including carrion.[5] They are sometimes called the 'undertaker bird' because of their striking height of one and a half metres (around five feet) and their large, dark, cloak-like wings that span up to three metres (around ten feet), and some people do indeed find them intimidating. Yet others, like the award-winning photojournalist Curity Ogada, marvel at their social behaviour, and admire how they groom one another and keep their white underbelly feathers clean in a demonstration of surprising gentleness that belies their imposing presence.[6] And it is this cleaning action that benefits the birds themselves and all city residents.

Marabou storks often follow vultures in their scavenging quests, and there is much we can learn from what happened in India when humans failed to be good neighbours to vultures. When I visited New Delhi in the 1990s, I was struck by the sight of cows wandering the roads and vultures perched above, both echoes of Nairobi. Same, same but different.* Little did I know then that these cows and vultures would become fatally linked.

In their research on the near collapse of India's vulture population between 2000 and 2005, Anant Sudarshan and Eyal Frank discovered that farmers had been using a cheap painkiller called diclofenac to treat their livestock. When the vultures consumed the carcasses of these cows, the medication, previously used only for humans, caused kidney failure in the birds. The result was a catastrophic loss of vultures, a population that had consumed millions of carcasses each year.

But the tragedy did not stop there. In communities that experienced the greatest loss of vultures there was also a significant increase in human deaths – an estimated 104,000 additional deaths annually, amounting to half a million over five years. Diseases that vultures had suppressed through their critical role as nature's sanitation engineers spread unchecked in their absence. By consuming animal carcasses, the birds prevented these from being scavenged by feral dogs, known vectors for rabies. Their actions also helped avert the contamination of water sources that could occur if the carcasses were left to decay. Although the use of diclofenac in livestock was eventually banned after the link was discovered, it will take years, if not decades, for India's vulture population to recover.[7]

---

* Same, same but different is a Kenyan English expression.

Vultures are often described as a keystone species – organisms essential to an entire ecosystem, whose absence can trigger widespread ecological disruption. We typically think of keystone species in terms of what they actively provide, such as honeybees, whose pollination sustains both plant life and human agriculture. But we often overlook species such as vultures, whose critical contribution lies in what they remove, helping us manage waste and maintain healthier environments. Precisely because we misunderstand or undervalue these interactions, we tend only to recognize their importance through the devastating consequences of their absence, rather than proactively appreciating and safeguarding their presence.

*

I was taught to pay close attention to nature in cities by a man who was a practical naturalist. Mzee Warũingi was the local mole-catcher.* He made a living in Karen – an area that was rapidly 'being tamed'. As more people in Karen fenced plots of land, cultivated lawns and built homes forming new communities, so his varied set of skills became highly valued. The fashion for immaculate gardens – lush lawns bordered by roses, exotic plants and a few indigenous flowers – meant that no one wanted their perfect carpets of grass marred by the small mounds of soil that moles left behind.

I imagine that Mzee Warũingi would have been in his late fifties or early sixties when he first entered my family's life. As a child aged somewhere between five and eight, I remember that he seemed ancient to me, and remained resolutely ancient until he passed away when I was in my twenties. In my imagination, he never gained any more wrinkles than those already etched into his face when I first met him. A short, slim

---

* Mzee is a Swahili term meaning 'elder' or 'old man'. It is also a respectful form of address for someone older or of high social status. In East Africa, especially Kenya, Uganda and Tanzania, it is commonly used both affectionately and respectfully.

and sprightly man, his features appeared moulded from the same clay that surely created him; he seemed at one with the earth. Although he wore a European-style jacket and trousers, he also adorned himself with a traditional Kikũyũ headdress of woolly sheepskin, complete with an ostrich feather. Wherever he went, he carried the tools of his trade: a *gunia* slung over his back, and, in his right hand, a *rungu*.* His hands were rough, the soles of his feet resembled distressed leather, and it was clear that he spent most of his life outdoors. Not so long ago, spending most of the day outside one's house wouldn't have seemed unusual at all. Many people in Kenya still live this way – their homes primarily places for sleeping, while daily activities like herding animals or tending crops all unfold outdoors in nature. But in a community rapidly charting its course towards a modern urbanity, Mzee Warũingi appeared displaced, and in reality, he was.

Some from his Kikũyũ community had been forced off their land by colonists, while others were rounded up into concentration camps, and tortured and murdered for resisting occupying forces. Mzee Warũingi lived within this complex postcolonial condition: unable to return to the life of his ancestors yet armed with an exquisite knowledge of nature that allowed him to sustain himself, albeit precariously, in a swiftly changing world. As my family kept an open house, he was treated just like any other

---

* A gunia is a sturdy versatile coarse hessian sack used across East Africa for storing and transporting agricultural goods like maize, beans, potatoes, charcoal and other produce.

A rungu is a traditional East African wooden club, often used symbolically or ceremonially, though historically it was also employed as a weapon. It is carved from hardwood with a smooth handle and a rounded, heavy head. In Kenya, especially among Maasai, Samburu and Kikũyũ communities, rungus carry cultural significance as symbols of authority, leadership and respect. Elders, community leaders or warriors traditionally carried them as emblems of power or status.

guest – though no matter how many times my mother invited him in, he preferred to take his tea outside.

Because he was an expert mole-catcher, he had access to well-to-do households in Karen whose occupants were eager to rid their manicured gardens of these 'infernal pests'. I must admit, though, that as a child I liked those little mounds. Just as I believed tiny people lived in the grass, I was equally certain that others must live underground, caring for the moles who, as Mzee Warũingi explained, don't see very well. He was exactly the sort of person that a curious child, eager to learn about the world outside of books, would gravitate towards.

I was especially taken by Mzee Warũingi because he helped me to understand the world I was actually living in. I grew up on a diet of British and American children's books, which, while fascinating, described a kind of nature that didn't exist outside my window. When Enid Blyton's characters set off on adventures, they wandered through forests and countryside that felt foreign. They were landscapes I could only recognize when we travelled to the UK. Closer to home, books that did describe Kenyan nature were mostly field guides: well-researched, full of information, but written with adults in mind. They were reference tools, not invitations to explore or play. I pored over them anyway, eager to learn, but I often felt that I was translating knowledge not written for me.

Mzee Warũingi was different. He took my stories seriously – the ones about the people I was certain lived in the mabingobingo grass. He would listen, ask me to describe them, and enquire about their lives. He said there were always two worlds: the one we could see and the one that was invisible, and sometimes they crossed over. As I grew older, I realized that he wasn't just talking about the material world and the world of imagination, but about how the imaginative could shape the material in real and lasting ways. While it was my travels with my parents that led me to become an anthropologist, it was Mzee Warũingi's lessons on nature that paved the way for me to become an anthropologist of the imagination. What we are able to imagine directly shapes the worlds we create – and it also determines what we are able, or unable, to see. Paying attention to the nature we live with in cities is a critical part of this process. And it was Mzee Warũingi who taught me, as he put it, 'how to read the birds'. His use of the word 'read' was deliberate – he used to say that one could read nature just like people read books.

By the time I conducted my first informal case study on ibises in Nairobi – nothing official, just a field diary where I noted their behaviour, spoke with people about them, and tried to make sense of what I was seeing compared to what the field guides reported – Mzee Warũingi had already died. I often wonder what he would have made of his apprentice. When he was preparing for his death, he even taught me the very skill that had sustained him – how to catch moles. One last time, he took me into the family compound where we had spent so many hours together in my childhood and swore me to secrecy. Then, for the first time, he opened his gunia and revealed all of its contents. On a mild *kipupwe* (cold season) afternoon, he gave me his final lesson.

Because I am sworn to secrecy, I can't teach you how to catch moles, but I can share Mzee Warũingi's method for reading nature. I went on to 'read' the hadada ibis (*Bostrychia hagedash*) – a large, iridescent-green bird with a long, curved bill and a famously raucous call. Once more commonly found in wetlands and rural areas, the hadada has become a familiar presence in Nairobi, where its loud, laughing cries echo through the early morning streets. I began to observe how it had adapted to city life, and slowly came to understand how it had made Nairobi its home, where it is now a common sight, and sound. There were not many ibises around when I was a child even though Nairobi's unique climate makes it home to over 677 species of birds.[8] Despite being near the equator, high-altitude cities like Nairobi boast markedly cooler temperatures than lowland equatorial areas. This is due to the thinner atmosphere at higher elevations, which provides less insulation, creating a more moderate climate compared to the intense heat often associated with equatorial regions. These cooler temperatures, paired with the consistent solar exposure of the tropics, form a distinct and stable climate ideal for many specialized ecosystems. Or, as I am known to say, 'Nairobi has the best weather in the world.'

Nairobi has average temperatures ranging from about 10°C (50°F) at night to 26°C (79°F) during the day. The temperature can vary slightly depending on the season, but it generally remains moderate throughout the year. This (alongside the city's abundant food supplies for wildlife) supports black kites and many other types of birds. There is also a reason that Nairobi is known as the 'green city in the sun', as its special environment means that it gets decent amounts of rain but not in a uniform pattern. The result is a collection of microclimates leading to a patchwork of complex environments that can support a wide range of plant and animal species right in the middle of a city.

What Mzee Warũingi taught me was that understanding these microclimates was key to unlocking what forms of nature would thrive there. Humans, often without realizing it, actively shape these microclimates, such as the lush, carefully maintained lawns in Karen with which Mzee Warũingi himself was involved. Largely oblivious to their role in creating a welcoming environment for certain species, many residents soon found themselves awakened just before dawn by an alarm clock they hadn't set, the loud and unmistakable *haa-haa-haa-de-dah* calls of the hadada ibis, an unintended consequence of their own landscaping choices.

As the perfectly manicured lawns and other hallmarks of modern urban living spread across Nairobi, so too did the hadada ibis. Once a bird of wetlands and woodlands, it quickly adapted to city life, drawn in by the very landscapes people had so carefully cultivated – soft, irrigated lawns, garden ponds, swimming pools and tree-lined streets that offered both food and shelter. Unknowingly, humans had created the perfect habitat, and the ibis took full advantage.

But the hadada ibis is more than just an uninvited wake-up call; it reminds us that nature will thrive in the most unexpected places. In South Africa, it was once considered a threatened species, heavily hunted to the point of requiring legal protection. But instead of fading into obscurity, it adapted to human-altered landscapes in ways that no one had predicted. As wetlands and forests were degraded, it found new opportunities – on farms, in city parks, even on the edges of airport runways. The very environments that had pushed other species out became its refuge, proving that conservation doesn't always happen in distant, untouched wilderness. Sometimes, it happens right under our noses, in the middle of our cities.

Ibises don't just survive in urban spaces, they flourish, and in doing so, they bring their own benefits. By probing the soil in search of food, they aerate it, improving water absorption and enriching it with nutrients. They provide natural pest control, clearing out snails, insects and larvae that would otherwise feast on gardens and crops. Yet despite their ecological usefulness, many still see them as a nuisance – mostly because they refuse to respect human sleep schedules. Their calls are not just loud, but deliberate, adjusted to rise above the hum of traffic and urban noise, ringing out before the streets come alive and again, in the evening, when the city starts to settle.

Even their breeding cycles have shifted to match the rainfall patterns of the city, another sign of how deeply they have embedded themselves into this human-altered landscape. If Mzee Warũingi were alive, I have no doubt that he would be able to predict the rains simply by watching them. But, as his lowly apprentice, I still have much to learn. What I do know is that the hadada ibis is a reminder that the divide between the wild and the urban is not as clear-cut as we often assume. There is more to cities than being places where nature is erased – they can also become unexpected sanctuaries, where the rhythms of the natural world continue, even if they wake us up a little earlier than we'd like.

Digging deeper into their cultural meaning, I learned that the ibis and its relatives hold profound symbolic meanings that stretch across cultures and centuries. It seems that humans have always sensed something special about these birds, recognizing them as carriers of wisdom, resilience and protection.

In ancient Egyptian mythology, for instance, the ibis was revered as the earthly embodiment of Thoth, the god of wisdom, writing and magic. Thoth himself was often depicted with an ibis's head, reflecting the bird's sacred role as a guide between the worlds of the living and the dead. It was Thoth who maintained cosmic balance, intervened in conflicts among the gods, and taught humanity the skills of writing and knowledge. Even the ibis's practical behaviour, feeding on snakes and pests, was translated symbolically into myth. The Egyptians believed ibises warded off the winged serpents that flew annually into Egypt from Arabia each spring, thus protecting communities from harm. It is striking to think that this ancient association might have found its echo, however faintly, in the gardens of Nairobi, where hadada ibises similarly serve as natural pest controllers.

Beyond Egypt, the ibis appears again and again in cultures worldwide, and with similar connotations of protection and insight. In some Christian traditions, it symbolizes honesty, justice and arbitration, and is celebrated for its perceived antagonism towards serpents and therefore deceit. Greek mythology, too, demonstrates a connection to the ibis through Hermes, the messenger god who guided souls to the underworld, an analogous figure to Egypt's Thoth.

Even in a quite different part of the world, in Native American myths from the Gulf Coast, the ibis is celebrated as a symbol of resilience, and is connected specifically with storms and hurricanes. Folklore describes it as the last bird to seek shelter before a storm, proof of its intuition and ability to adapt – qualities I now clearly discern in the urban hadada ibis.

These stories demonstrate that humans have long recognized and revered birds such as the ibis precisely for their adaptability and intelligence, and for their role as connectors – between people, places and even worlds. Today, in our urbanized lives, we often forget about these ancient bonds. Yet, standing in Nairobi, hearing the distinctive cry of the hadada ibis above the traffic noise, I wonder if we are still, in some deep and instinctive way, listening for wisdom from these ancient guides, as they (not-so) quietly teach us how to live better alongside nature.

*

Yet the presence of birds like the hadada ibis reminds us that living well with nature in cities is never just about individual encounters. Every bird call we hear depends on contested spaces of land, water and air. Urban land carries multiple and often competing layers of value, and with rising real estate costs – here as elsewhere – nature is too often reduced to economic potential. The ibis survives because wetlands, trees and open ground survive. Protecting those spaces requires more than private wonder; it demands public resistance. We need examples where one person's care for a patch of earth, a wetland or even a single tree sparks others to reimagine what a city can hold, and for whom. Those sparks of resistance are what transform fleeting moments of connection with nature into the collective will to defend it.

One citizen, more than any other, quite literally put her life on the line for Nairobi's trees. In 1977, Wangari Maathai, a Kenyan biologist and environmentalist, took what seemed a simple action: she planted seven trees in Kamukunji Park, each one representing a traditional leader who had resisted colonial rule. This act of remembrance and resistance grew into the Green Belt Movement, which mobilized thousands of women across Kenya to plant trees as a form of environmental restoration and political protest. In 1988, when she went to plant a tree in a protected forest on the edge of Nairobi – a forest under threat from corrupt politicians illegally parcelling it out for personal gain – she was met with brutal opposition. Around two hundred hired thugs armed with machetes, bows, arrows, whips, clubs and stones accosted her and her supporters. Her non-violent action left her hospitalized but cemented a powerful idea: that a sustainable environment could not exist without democracy. For Maathai, trees were more than nature – they were about dignity, sovereignty and justice. Her work challenged both ecological degradation and authoritarian power, and it transformed how many Kenyans saw the value of land. In 2004, she became the first African woman to receive the Nobel Peace Prize.[9]

Wangari Maathai often referred to herself as a hummingbird. And, many years later, whenever I took my daughter to walk in Karura Forest or any of the other parks that Wangari Maathai helped save, I would share both a story and a parable. First, I told her the story of Wangari Maathai herself. Then I shared the parable of the hummingbird that inspired Maathai.

The parable goes like this. A massive fire was engulfing the forest, and while all the animals stood by, watching the smoke and flames, a tiny hummingbird flew back and forth, carrying as much water as its little beak could hold. The other animals mocked the bird, disparaging its efforts, but the hummingbird carried on undeterred, saying simply, 'I'm doing the best that I can.' This is why Wangari named herself the Hummingbird – for doing her little bit. Within the parable lies a profound truth: that greatness is not a prerequisite for making a difference. For me, my 'little bit' has turned out to be about exploring the possibilities already present in cities and finding ways to share how to strengthen them. And though I wouldn't call myself an activist, I believe that we all have a role to play in protecting and positively engaging with nature.

The idea of doing 'my little bit' – of making sense of how human-altered landscapes can still be havens for wildlife – has been with me for as long as I can remember, though I didn't have the language for it as a child. Back then, I simply felt the pull of two seemingly opposing forces. On one side was Mzee Warũingi, unwavering in his belief that learning to read nature was *the* way to live. Yet his own community's history was proof of how contested land could become, how people could lose everything – sometimes even their lives – over their connection to it. On the other side was the stark reality of what was happening to people like Wangari Maathai, who faced brutal resistance simply for insisting that trees, forests and green spaces mattered. And yet, in the backdrop to it all, nature itself continued, perpetually adapting, healing and reshaping the world in ways both subtle and dramatic. The question that stayed with me was how to hold all these tensions together – to acknowledge the risks, the history and the struggle, while still finding ways to act, to contribute something meaningful. If nature itself refuses to give up, constantly finding ways to survive and regenerate, then perhaps we, too, can learn from it. We can learn how to live with the natural world and insist on shaping a future where both people and nature can thrive.

Maybe that's why I've always been drawn to the small, stubborn acts – the planting of something where nothing is meant to grow, the quiet claiming of a forgotten corner of the city. If nature keeps reaching for light through cracks in the pavement, then so can we. Whether it's a borrowed plot on the edge of a railway line or a seedling smuggled into a patch of municipal scrub, these gestures feel like replies. They may not be grand solutions, but they are critical responses all the same. Ways of saying: I'm still here. I see this land. I want to tend it.

Chapter Five

Grown in the Soil: Cultivating in Crisis

I get home from school with Sister's seeds zipped into the inner pocket of my school bag, wrapped tightly in a corner of my hanky so no one will suspect a thing. Bean and carrot. She gave them to us after morning prayers – said it's good for girls to know how to grow things. I nodded solemnly like everyone else, but inside, I'm buzzing.

I wait until the house is full of voices and Oloo is busy getting the dogs' food ready before I slip out of the kitchen door and round to the back of the house. I already know where I'm going. There's a space under the yesterday, today and tomorrow bush (*Brunfelsia pauciflora*), just to the right of the veranda. It's shady and quiet and smells a little like stories. It's not part of the official vegetable garden – too close to the house, too small to matter – and that's exactly why I love it.

I dig a little with my hands, checking over my shoulder now and then. Then, one by one, I tuck the seeds into the soil. Not too deep, not too shallow. I follow Sister's instructions to the letter. I say a small something to them – not quite a prayer, not quite a poem – then pat the earth flat again like nothing ever happened.

For the next few weeks, I keep visiting. Sometimes twice or even three times a day. Quietly. Carefully. I have to take care when I water them as someone might notice but I whisper to them every time I do. And it works. One afternoon, I see it: tiny shoots poking up through the soil like they're waving. I kneel down and just stare. They're so small it feels like magic that they're even real. I brush the soil gently away from their stems and smile at them like I know a secret.

The days go by and the shoots get bigger – first with leaves, then the hint of something swelling beneath the earth. Eventually, I feel it before I see it: a bulge in the soil, the promise of something orange. And one afternoon, when I can't wait any longer, I crouch down, press my fingers in, and gently tug.

It comes loose with a soft pop, roots and all. A real carrot. Small, slightly crooked and absolutely perfect. I shake off the soil and bite in – no washing, just that raw crunch and sweetness. It's the best thing I've ever tasted. No one told me carrots could taste like this. I sit here chewing slowly, as if the taste will disappear if I rush it.

That night, I can't sleep. I've grown something. All on my own.

But a few days later, it happens.

I'm crouched beside the bush, checking the leaves, wondering when the beans might be ready. I take a deep breath, and then another, but something's off. A scratch in my chest. A wheeze. My lungs start to tighten like they're being squeezed. The world blurs slightly, and I lean against the ground, waiting for the next breath to come.

Hours later, I'm wrapped in bed my lips still faintly tingling from the doctor's injections. The home-made steamer is whirring away next to me, working to keep my airways open. My parents are finishing off a conversation outside my bedroom door. I know what's coming before anyone says it. I can feel it in their tone of voice, both concerned but also relieved like they might have found the solution to a long-standing problem. And sure enough, a few mornings later, I walk outside, and the bush is gone.

Uprooted. Cleared. Like it was never there.

They didn't said anything about my secret garden, but it's gone too. The soil looks flat and blank. No little shoots. No wild green promise. Just empty earth where my garden used to be.

I don't cry. Not yet. I just stand there for a long time.

*

I'll never know if Sister knew the science behind what she was doing by handing out those carrot and bean seeds and teaching us how to tend to them. But decades of research now confirm what I felt then: growing something is good for us.[1] Really good. It can't be explained by the food itself, or even the flowers – it's about what happens inside us when we plant, cultivate and harvest. Studies show that gardening reduces stress, improves mood and increases our sense of agency and hope.[2] Despite my allergies, that feeling I had, as I crouched under the shrub checking for sprouts, wasn't just excitement. It was the quiet confidence that comes from doing something life-giving with your own two hands.

There's even evidence that certain microbes in soil – like *Mycobacterium vaccae* – can boost serotonin levels, acting like a natural antidepressant,[3] and that gardening builds resilience, especially in children, helping to develop patience, focus and a sense of responsibility. One recent meta-analysis found that time spent gardening has a measurable impact on mental well-being across all age groups, supporting emotional regulation, social connection and even cognitive performance.[4]

What I was experiencing, in the shade of that shrub, was an early practice in care, attention and belonging. That first carrot, pulled from the soil and eaten with the dirt still clinging to it, was more than a snack. It was proof that I could help something grow. And in times of uncertainty or illness, that knowledge – that life could still take root – felt like a kind of power.

My secret garden was just one of many types of gardens in Nairobi where people tap into that kind of power. All over the city, people cultivate even the tiniest scrap of land. When I was a child, *askaris* (security guards) stationed outside people's homes or blocks of flats would plant kale in small patches of ground near where they sat keeping guard over the properties to which they were assigned. Nairobi is full of these micro-farms. Even today, if you know where to look, there are mini farms that are hidden, tucked away near Nairobi's numerous streams and water courses. Drone pictures of Nairobi show that (from certain perspectives) Nairobi looks like a city planted on top of a collection of farms. Part of this is due to rural urban migration and people literally bringing their farms with them by way of seeds and plants. Another reason is the security that growing one's own food can provide in times of political upheaval. This need for self-reliance, quietly growing food in any space available, isn't just about tradition or habit. For many of us, it is tied to memory, shaped by the disruptions of history.

My concerns about food and food security were inextricably linked to times of political upheaval in Kenya. In August 1982, when I was nearly seven years old, an attempted *coup d'état* took place. Although Daniel arap Moi (the erstwhile dictatorial Kenyan president) quelled the coup relatively swiftly, the measures put in place were a shocking reminder for Kenyans such as my parents who had grown up in the colonial era. In those days, curfews (forbidding people to be outside their homes at certain times of day) were used as a means of collective punishment of Kenyan Africans, the effects of which were far-reaching, including the destabilization of food security and resultant malnutrition and death. The 1982 curfews left indelible memories for the Nairobians who lived through them. Residents were first advised to stay at home until the trouble was over, then an indefinite curfew followed, banning them from being outside their homes between 6pm and 7am on threat of arrest. While this period lasted just over a month, my memories of it are heightened because, right at the start, my father accidentally broke the curfew, and we had to race home through deserted city streets. I had never seen the city like it; not a soul was in sight. In the quiet that followed, my mother turned her attention to what she knew best: preparing for uncertainty through cultivation.

My mother had grown up in a household that had hosted refugees from different parts of East Africa, including other parts of Kenya. One of the ways in which my grandparents' household had been able to support numerous people over decades is the wide variety of crops they grew, the animals they reared and the fish they farmed. So my mother instructed Oloo that we had to intensify planting for upcoming *vuli* (short rainy season) because we had no idea how long the curfew would last. Until I began to farm in response to crisis, I had not been aware of this inheritance: the mentality that one should, ideally, always be growing food, and that this production must be intensified during difficult times.

In the decades after the attempted coup, several factors coalesced to make urban gardening a much less practised activity than it had been in my childhood. The most striking was the population density increase – between 1979 and 1999 Nairobi's population and concurrent population density increased by more than two and a half times. As in much of the rest of the world, urban land was at a premium, so opportunities for guerrilla gardening shrank as the population grew, with more people arriving to live in Nairobi and people being displaced either through corrupt administrations or because of the legacy of colonial land laws superimposed onto traditional land systems of management. When post-election violence occurred in 2008, I responded to it viscerally, with memories of the intensification of gardening that had occurred not just on the family compound but among our neighbours. When violence erupted, it was as though decades of pent-up frustration and historical grievances exploded into chaos. The incumbent president, Mwai Kibaki, was accused by his challenger Raila Odinga of vote-rigging, setting off widespread protests and clashes.

The violence revealed how fragile Kenya's democracy still was, and how deep the scars of our colonial history ran. Nairobi's geography still bore the imprint of colonial-era racial segregation, not unlike apartheid-era South Africa: leafy, spacious, former settler suburbs like Karen – where I grew up, surrounded by large gardens – stood in stark contrast to overcrowded, under-serviced neighbourhoods like Kibera, where much of the city's population lived on just a fraction of its land. According to estimates at the time, 60 per cent of Nairobi's population lived on less than 5 per cent of its land, while over 80 per cent had no formal access to urban land at all.[5] Land has always been political in Kenya. During colonial rule, Africans were forcibly removed from fertile areas and denied the right to grow certain crops. Even after independence, land distribution remained grossly unequal, and those inequalities were often mapped onto ethnic lines, deepening mistrust. The election didn't create those tensions – it simply cracked the surface.

This stark inequality had its roots in the colonial government's policies, which were designed explicitly to limit and control the African Black population in urban spaces. The echoes of Charles Eliot, Kenya's first colonial governor, who famously declared Kenya 'a white man's country', still reverberated through Nairobi's urban planning and land distribution decades after independence.[6]

It was in this context of crisis, echoing past experiences yet facing unprecedented new challenges, that I started an organic farm in Kibera. This was a way of creating a green sanctuary of resilience within Nairobi's crowded urban landscape. As my decision to do so solidified, I turned down offers from friends in the UK to help fly me and my family out of

Kenya to safety, asking them instead if they would they fund the farm. Everyone agreed.

*

What does it take to start a farm? For most people, their imagination immediately takes them to the countryside, to acres of fields that may or may not host animals as well as crops. Because I grew up in a neighbourhood where people routinely treated part of their gardens as farms, I would argue that a farm can be anywhere. All that is required is access to some land, some seeds and a willing group of people to tend to them. The appropriate weather helps, particularly if coupled with the patience to let nature take its course.

In the case of our farm in Kibera, what we had at first was just about the size of a football field. It was a waste-strewn strip of land by the railway line, one of those places the city forgets as the rubbish piles up. Yet, for that very reason, it was accessible. No one had claimed it. Or rather, no one had cared enough to. And sometimes, that's exactly the kind of place where something transformative can begin. I had no grand designs, only a simple impulse – an echo of what I had inherited from my mother, and from her father before her: in times of uncertainty, cultivate. What I hadn't yet realized was how deeply this act of cultivation would challenge the colonial inheritance of what it means to belong to a place, and who gets to care for the land.

The people who gathered around that sliver of land in Kibera weren't strangers to reinvention. Many were young men who described themselves as 'self-rehabilitated ex-offenders', a phrase I'd never encountered before, but one that made perfect sense. The phrase held recovery within it, and something else – agency. These were people who had already made the decision to change their lives, long before any project came along to help them do so.

One of them told me, 'I looked around and realized I didn't like coming from a dangerous place. Then I realized the danger was coming from me.' The group was called Youth Reform. They didn't need saving. What they needed was space to build something of their own. Many were people of faith, and their devotion brought back memories of Sister's lesson on the importance of growing something. Youth Reform members are mostly Muslim, I am Catholic, but none of us felt that difference as we worked together. In a garden, there is always more than one story unfolding.

> *I made a garden for God.*
> *No, do not misunderstand me*
> *It was not on some lovely estate or even in a pretty suburb.*
> *I made a garden for God*
> *in the slum of my heart*
> *a sunless space between grimy walls*
> *the reek of cabbage water in the air refuse*
> *strewn on the cracked asphalt ...*
> *the ground of my garden!*[7]

These words, written by Ruth Burrows, a cloistered nun on another continent, echoed what we were doing in that open patch of land. Youth Reform began clearing. We fished out old batteries, plastic bags, broken bits of corrugated iron from layers of decomposing refuse. We knew we couldn't remove it all – not safely, not without harming someone else. So we covered what we couldn't clear with a net and left it where it was, beside the garden. A kind of scar. A reminder of where the land had come from, and what it was working to leave behind.

The batteries, however, worried me. Although we were working to be able to till the land before the long masika rains of March and April, I didn't think it would be appropriate to grow food on what could be polluted soil.

A childhood friend of mine, now a skilled organic farmer, began teaching the group how to become organic farmers themselves. But I wasn't convinced that the soil was safe. I started looking for experts to test it and help us figure out what to do if my suspicions were confirmed.

The results of the soil testing turned out to be exactly as I'd expected. It had high levels of zinc, probably from those discarded batteries we had found during the clear-out. The potential of this soil was not going to be straightforward to access. I was struck by the ways in which slums that get labelled as problematic are often at the receiving end of problems created by others. It turned out that there were two soils: the darker earth that naturally belonged in Kibera, and a redder, lighter soil that had been illegally dumped by those carrying out construction in another part of the city. It was possible to trace the Kibera soil because, despite all the rubbish heaped on it, there were some plants growing underneath. When Moha, the social secretary of Youth Reform, asked some elders about these plants, it brought back memories of the days when Kibera was not a slum but a collection of villages where agroforestry was carried out. Moha and the other Youth Reform members are Nubian, a community that was resettled in Kibera by the British colonial government following their services to the King's African Rifles during both the First and the Second World Wars. Kibera's name in the Nubian language is Kibra, which means forest. But there were yet more surprises to be uncovered by the simple act of listening to what the land was working to reveal.

First, though, we had to heal the soil. The experts we consulted agreed to teach us how to work with plants in order to do so. We learned that there is a specific category of plants called phytoremediators that are able to remove, break down, or stabilize harmful pollutants from the soil. Sunflowers, for example, are excellent at removing zinc. They became the 'pioneers' of the newly established farm. Something else happened that we did not expect. As the sunflowers grew – their bright heads hard to ignore – the farm became a meeting place as people were drawn by their beauty. The flowers became our calling card, a symbol that something new was taking root, and their golden faces began to shift the way people moved around that space. People slowed down to look. Children came to ask questions. Conversations started. The local communities wanted to know what was going on – on what used to be a rubbish dump. We learned that one female elder knew how to make cold-pressed sunflower oil, so, when the fresh batch of sunflowers grew they would have a practical use as well as an aesthetic one. In the meantime, the sunflowers that were quietly working hard to de-toxify the soil stood tall and unapologetically hopeful.

Once the soil was loosened and we'd added compost, we began to plant food. Leafy greens came first: sukuma wiki, spinach, terere (amaranth). Then tomatoes and onions. The staples that Nairobians knew, that tasted of home. Crops that could feed a household, that would sell quickly at market. Crops that, grown well, could convince others that this wasn't a novelty, but a legitimate farm.

We didn't even have proper tools at first. But we had commitment and goodwill that travelled with the donations that came in via word of mouth from around the world, and as the plants began to grow, so too did the sense of collective ownership. One of the young men constructed a makeshift greenhouse out of discarded polythene sheets and wire. Another began teaching younger children about seed spacing and how to check for pests. Before long, the site began to function not only as a farm but as a space of learning, of recovery, of reimagining. At its peak, the farm was feeding 74 families. More than that, it birthed a clutch of small businesses, and as Kenya returned to greater security, visitors from around the world began to visit and invite Moha and others to countries as far away as Brazil to share their knowledge and experience. A group of farmers from Wales struggling with intergenerational handover on their own farms volunteered for a time on Youth Reform's farm and returned to Wales revived, and with new ideas to implement in their own situations. The University of Nairobi donated chickens and rabbits to the farm, and students carried out placements there. The story of the farm drew a significant following.

When I speak about it now, I notice how the questions I get are usually about the violence that preceded the farm's establishment, or the people involved, sometimes from a place of curiosity, but often from one of disbelief, as though growing food in a slum is inherently improbable. But to me, what was improbable was not that it succeeded but that we ever stopped growing food in cities in the first place. We forget that all cities at some point in their history had sites where food was grown. Land, even the most marginal, holds memory. And if you look closely at any city, you'll still find those remnants: the fig tree that no one will cut down because it's believed to host spirits, the banana patch behind a block of flats, the sukuma wiki by the askari's hut. These are not accidents. They are ancestral whispers, reminders that land remembers who once cared for it.

In a way, that's what the farm became: a space where memory could be reactivated through action; where people who had been written off by the state and society could return to a kind of ancestral logic – cultivate, share, protect. Where we could collectively refuse the idea that only certain kinds of bodies in certain kinds of places deserve beauty or food or peace. Of course, it wasn't always easy. Some people who didn't fully understand the work of phytoremediation thought we were wasting time growing flowers that could not be consumed when we should just have been getting on with growing food. I was volunteering all my time, while juggling with breastfeeding a young baby and a return to work so I could earn some money. The political climate was also uneasy and it was difficult to know whether the tenuous peace would hold. But, through it all, we kept planting. And the act of planting – simple, repetitive, hopeful – became its own kind of resistance.

Looking back, I can see now that I was changed by that land. I had gone there thinking I was bringing something – knowledge, resources, maybe even healing. But what I received was far more profound. The farm in Kibera taught me that belonging is not a state you arrive at. It is a practice. A set of repeated actions that declare, 'I am here, and I care'. The quiet, radical statement it makes. Putting your hands into the soil is a way of claiming space without dominating it, instead forming a relationship with it. A recognition that to grow food is to declare that life can emerge even in places of precarity. And so, the farm in Kibera was never just about food. It was about rebuilding a sense of agency, of rootedness, of beauty in the face of brutality. It was, in every way, an extension of what my grandparents taught my mother, and what she passed on to me: in times of crisis, we plant. We find the forgotten places, we gather our seeds and we begin again.

*

Although I'd spent half my life in England, it wasn't until I found myself working on the organic farm in Kibra that my curiosity about a different kind of garden I'd often glimpsed near railway tracks throughout the UK was reignited. Allotments. These productive, often messy and anarchic spaces flash by from train windows, catching your eye just long enough to wonder what's growing there, and who's tending it. In my teens and twenties, I went wherever the train could take me and spent many hours watching for those fleeting green patches. Years later, back in the UK with a young family, and having decided to change career, those allotments, along with guerrilla gardens scattered across Oxford, felt like the perfect place to root my doctoral research.

Allotments and railway tracks have been linked for over a century, a legacy of both practicality and social need. As Britain industrialized in the 19th century, railway companies acquired long ribbons of land that bordered the tracks, spaces too narrow or awkward to build on, but perfect for growing food.[8] Initially, these plots were offered to railway workers as a modest but meaningful perk. Later, during both world wars, the relationship between land and survival took on new urgency: under campaigns like 'Dig for Victory', even railway embankments and sidings were transformed into sites of resilience, growing food for a nation under pressure.[9] Legislation like the Cultivation of Land Act 1916 made it possible to requisition so-called idle land for allotment use, turning marginal space into vital ground. That legacy still lingers today.[10] And those small, stubborn rectangles of green, just seen through a train window, tell a quiet story of collective effort and everyday endurance.

Allotments, those small plots rented out by councils for people to grow their own food, are more than just bits of land. They're a cherished part of British urban life, especially in times of hardship or recovery. My research explored how people connect to land and to each other in these green patches, in some of the most densely populated parts of Britain. Part of my fieldwork took place during the years of the COVID-19 lockdowns so when it was announced that allotmenteering would count as one of our permitted hours of daily exercise under the UK's lockdown regulations I literally wept with joy and relief. The allotments I had come to know intimately would remain essential sanctuaries, not just for me, but for the communities that tended them.

It was around this time that Sam Skinner, an artist involved in local eco-art initiatives, offered me an allotment plot to use as an outdoor studio. He said I could do anything I wanted with it and that he would help me. I accepted without hesitation. There was a gentleness in the offer – a gesture of trust and generosity that seemed entirely in tune with the moment. I immediately knew what I wanted to do. As a historian of the First World War, I had long been fascinated by how people worked with the land in the face of unimaginable devastation. I wanted to recreate a First World War-era allotment – a 1918 plot, true to the time, planted with heritage seeds (traditional varieties passed down through generations, often more diverse and resilient) and guided by century-old gardening manuals.

It was an intuitive decision, but not a random one. Gardening, after all, has often emerged as a quiet form of resistance and repair during times of conflict and crisis, as I had experienced in Kenya in the wake of post-election violence. I instinctively felt that if we were to work on the allotment, in the same way that people during the First World War had created their own allotments, we would learn something and that it would be a sustaining practice. And although Britain was not at war, the newspapers during this period were very much using wartime language to describe the battle against the new virus. There were also echoes from the past that were not going unnoticed, with comparisons being made between the 1918–19 so-called Spanish flu pandemic and COVID-19. What I observed, and what I experienced myself, was a kind of temporal folding: in 2020, people were gardening not only for food and exercise but to anchor themselves in something deeper – in memory, in meaning, in a connection with those who had gardened before us during similarly uncertain times. Some corners of the internet itself turned into a giant garden, with people showing off photos of seedlings they had managed to germinate for the first time and making use of every space they could, from balconies to slivers of urban gardens.

My work as a First World War historian, alongside my anthropological research, gave me a sense that there was something deeply resonant in returning to that historical period – a time shaped by war, scarcity and the Spanish flu pandemic. As I trawled through newspaper archives from the period, I was struck by how vividly present these concerns were to people then, and how clearly allotments were framed as part of the solution. One *Daily Mail* newspaper advert from 1918, for example, urged readers to take up allotmenteering, noting that gardeners were seen as healthier and therefore less likely to fall victim to the flu. My decision to take Sam up on his offer led to the creation of the '1918 Allotment', a public history project that became a place of reflection, remembrance and hope.

When I began researching urban allotments in Oxford before COVID-19, I was surprised to find that there wasn't even a city-wide waiting list. Plots were relatively easy to come by then. Depending on where you lived, and if you didn't mind a bit of a journey, you could secure a lease for a plot on the very same day you turned up at a site. However, once the pandemic began, demand surged dramatically, and waiting lists rapidly formed across Oxford as people sought solace and sustenance in gardening. And this wasn't the first time allotments had been a vital sanctuary during a national crisis. Allotments had sustained Britain during the First and Second World Wars, the 1970s oil crisis and austerity. The list is long.[11]

*

The 1918 Allotment began as a living experiment: a fully organic plot, cultivated with the same heritage seeds and hand tools allotmenteers might have used over a century ago. Planting followed instructions from T W Sanders's *Kitchen Garden and Allotment* – a guidebook first published in 1918. At first, the work seemed like quiet homage. But as the first shoots pushed through, something shifted. The past did not stay in the past. It came alive in the textures of the soil, in the heft of a beetroot, in the lift of a spade. History was no longer abstract; it lodged beneath fingernails. Beans were supported using 1918 methods – coppiced sticks, angled into A-frames and tied with twine. Strawberries that had survived the gap between different allotmenteers on the same plot were turned into jam and made ready for visitors.

As the lockdowns eased, visitors began arriving. Some stayed to chat, others dug silently. The project increasingly revealed the agency of the plants themselves. The heritage seeds that were sown – cabbage, beetroot, onion, carrot, kale, King Edward potatoes – had their own rhythms and their own resilience. Some crops grew beyond expectation, but one morning, before an event, we found that blight had taken out all of the tomatoes. Sam rushed to fill the gaps from the nursery bed of seedlings. Such replacement was possible in 2020, but would have been far harder in wartime, when everything grown had been a matter of life and death.

Insects, too, played their part. They crossed human boundaries without care for fences or signs. Bees buzzed through the plot, pollinating as they went. Their presence was a reminder of how closely human survival is bound to theirs. In 1918, pesticides had been hailed as modern miracles, and films like *Allotment Holder's Enemies* (1918) portrayed the extermination of insects using the language of battle and war. And yet a century later, many allotmenteers planted pollinator-friendly flowers in recognition of how vital and vulnerable these more-than-human allies are. Some of the ornamental blooms cultivated here became, in their own understated way, memorials to insects lost in the wake of industrial warfare.

There was a haunting resonance between the destruction wrought by war and the ways in which the land has been treated since. Lines from First World War poetry found unsettling echoes in the garden's literature – the language of control, domination, eradication. Even T W Sanders's practical advice on cultivating beets or applying blood meal began to flicker with other meanings. His instructions seemed almost like a poem: 'thrust / the blade down deeply / press back the handle / twist.' It could have been about gardening. It could have been about bayonets. The ambiguity sat heavily.

Over time, visitors began to bring their own stories. One woman travelled across the country with her son to spend the anniversary of her father's death in this space, because he had been a Second World War allotmenteer. Another came for poetry and unearthed the truth about her grandfather, who had fought (and survived) on the East African front during the First World War, something she had never known. These stories grafted themselves onto the allotment like second roots. What had started as research quietly became remembrance.

The allotment went on to offer something more than fresh vegetables. It became a space for grief, for imagination, for reconnection.

Public events were hosted there – readings, performances, shared meals. Poets wrote and performed among the beds. A forest-bathing session was held in the trees bordering the site. Rum was toasted, just as soldiers once did before going into battle. Visitors described the experience as 'multi-sensory', 'hands-on', 'healing'. One page of the guestbook was left streaked with soil, a quiet marker of presence. At a time when touch felt dangerous, the allotment recalled Octavia Butler's words: 'All that you touch you Change. All that you Change, Changes you. The only lasting truth is Change. God is Change.'[12]

The 1918 Allotment was not a traditional memorial. It did not ask visitors to read a plaque or remember a date. It asked them to notice: the curve of a potato, the shape of grief, the intimacy of working with soil. In doing so, it also became a counter-memorial – a gentle act of resistance against historical silence. The 1918–19 flu pandemic took millions of lives but left few monuments. Why? Perhaps because it was not a war. Perhaps because there were no heroes. But standing in that hundred-year-old allotment, growing kale and memory side by side, the absence of commemoration felt stark. Here, that silence began to be answered – not with statues, but with strawberries.

*

One of the gardens I cultivate in Oxford is a community garden in the shadow of a University college building. A group of us who like to grow, some of whom have their own gardens but are here just as much for the company as the cabbages, get together on a Sunday to tend to it. We also relish the fact that this is prime urban land. Its value is usually measured in pounds per square foot, but here we're inscribing a different kind of worth; a worth that is beyond what real estate can offer, one tied to values of care, cultivation and community. One of the growers is a fellow researcher and I ask him why growing on urban land matters so much to him. He says that all of his urban life feels disconnected from reality. Sitting in archives, writing papers – even though it financially supports his family – doesn't bring the same deep satisfaction as when he literally puts food on their plates. He feels that urbanity has stripped him of a network of relationships that nature is active in and that feeds his soul.

Later that day, as I look at a grainy image of wartime allotmenteers, I am struck by the plants that bound their plots, characterized as they are by tidy rows of vegetables – these plants are flowers. When every inch of space mattered, when the country was at war, when food was scarce, room was still made for beauty. I think of my fellow grower's words, and how the way we are nourished matters just as much as the nourishment itself. It's not just about existing in cities that allow us to make a living. It is how we do so and what we choose to make room for that is just as important. Sister's words about taking care and growing come back to me in this moment. There is something radical about growing something, especially in places we've been taught to overlook, to pass by without a second glance. Whether in an informal settlement in Nairobi or the border of a railway in Oxford, gardens become places of return – to ourselves, to one another, to forgotten truths about who we are and who we might still become. In times of crisis, planting is more than an act of survival, it's a subtle rebellion.

It says: I am still here. I am still alive. I still believe in tomorrow.

Chapter Six

Held in the Soil: Plants, Memory and Resilience

I always wake up early – 4am is my usual time – so when the COVID-19 lockdowns begin, the stillness that unsettles so many feels, to me, same, same but different. There's something sacred about these pre-dawn hours, a hush the rest of the world seems, for once, to be tuning in to. As I walk outside, there's a stillness not just of quiet, but of absence: no footsteps, no traffic, no unexpected meetings. The city feels like it has escaped time.

And yet I'm never truly alone. This hour belongs to particular creatures. I glance into the park and sure enough, there's Patrick, shuffling gently along with his two dogs trotting at his side. He speaks to them in a low affectionate murmur, as if they are toddlers who need reassurance. A few minutes later, I spot Olivia the missionary, her voice rising clearly above the stillness as she sings hymns to the trees. There's something so unselfconsciously graceful about her singing that I find myself slowing down just to listen.

I scan the hedges of the other park across the road. I wonder if I'll see Jimi the fox this morning. She has a way of appearing without warning, pausing to look at me straight in the eye, as if in greeting, before slowly trotting off, as though we have just finished an important catch-up. Maybe Hannah the hedgehog will be out too, nosing along the curb. Bumble the badger sometimes ambles down the road alongside the park, her lolloping so endearing that it never fails to make me smile. And during lockdown we've gained a new companion: Dixie the muntjac, with her lopsided limp, weaving through the early light like she's on her way home after one too many.

These small, fleeting encounters stitch a kind of continuity into these suspended days.

I stop in my tracks when I spot a blackbird standing bold in the middle of the road, where it would normally never be. It's locked in a silent battle with a worm – tugging and twisting. Even the dogs pause, tilting their heads. The phrase 'the early bird catches the worm' lands differently now.

The worm is fighting back.

I don't know why I ever assumed the worm would just surrender. It doesn't. It writhes, resists, coils so fiercely that sometimes the blackbird drops it. Then, with a flick of its beak, the bird grabs hold again. They keep going – back and forth, a delicate, brutal tug of war. I don't intervene. I just stand there, watching, drawn in.

Eventually, I walk on. I don't need to see how it ends.

But within the ongoing silence, something else lingers – a sense of possibility. The kind that comes just before something begins. That's what I love most about waking up early: the way potential hangs in the air, undisturbed. You can only feel it when distractions are few, when the world hasn't quite remembered itself yet. In this pause, I can hear the breath a bird takes before it starts to sing. Even the faintest rustle in a bush carries weight, amplified by stillness, as if nature is clearing its throat before the day begins.

*

Work on the 1918 Allotment has me thinking about the aftermath. Not just about what comes after crisis, but how anything, or anyone, begins again. How do plants, nature and human beings not only survive upheaval but go on to grow, sometimes even flourish? The common poppy (*Papaver rhoeas*) is the first to raise this question.

> *In Flanders fields the poppies blow*
> *Between the crosses, row on row,*
> *That mark our place; and in the sky*
> *The larks, still bravely singing, fly*
> *Scarce heard amid the guns below.*
>
> *We are the Dead. Short days ago*
> *We lived, felt dawn, saw sunset glow,*
> *Loved and were loved, and now we lie,*
> *In Flanders fields.*
>
> *Take up our quarrel with the foe:*
> *To you from failing hands we throw*
> *The torch; be yours to hold it high.*
> *If ye break faith with us who die*
> *We shall not sleep, though poppies grow*
> *In Flanders fields.*[1]

The very flowers we associate with remembrance are what ecologists call ruderal species – plants that are specialists in disturbance.[2] They germinate fast, thrive in disrupted soil, and take root in unlikely places: where bombs have fallen, trenches have torn the ground, or the land has been stripped bare. These plants don't need rich soil or stability; they are adapted to adversity. In ecological succession, they're often the first to arrive and begin healing the ground, stabilizing it so that other, slower-growing plants can eventually follow. In ecology, succession is how landscapes heal, starting with tough, fast-growing species that move in first, make the soil liveable again, and set the stage for everything else to follow.

There's a tenderness in that resilience. Poppies are delicate, after all. And that's what strikes me most: how fragile the first signs of recovery can be. They don't arrive with grandeur or loud declarations. They can be easily blown about but they persist, even if they seem fleeting or easy to miss. These early signs matter. They remind us that beauty is still possible, even in broken places – and that's something we can lose track of, especially as humans, rushing forward.

There's another kind of wisdom in the way these plants operate. Nature doesn't presume that peace can be the default. It builds for disruption. It has whole species – ruderal, tenacious, adaptive – designed specifically for damage. In this way, nature prepares for crisis without fear. It sets up the next stage of regeneration even as destruction unfolds.

When we co-create gardens with nature – when we plant, compost, tend – we step into that cycle. We don't just recover from crisis; we prepare for the next one. The very act of cultivating a shared space becomes a rehearsal for future resilience. And the garden becomes more than a patch of soil – it's a quiet architecture of care, a structure for healing made from root, seed and bloom.

This is something I hold onto as we think about how cities rebuild. Regeneration isn't always about bricks and buildings. Sometimes it begins with something much smaller: a patch of disturbed earth, a handful of seeds, a poppy that blooms where nothing else will.

✻

Bethnal Green, situated in London's vibrant East End, exemplifies how urban spaces can evolve to sustain abundant biodiversity even amid human density and historical trauma. During the Second World War, the area endured extensive bombing due to its proximity to London's strategically vital docks and industrial sites, making it a primary target during the Blitz, which began on 7 September 1940.[3] The Luftwaffe's relentless attacks on the nearby docklands inevitably brought devastation to Bethnal Green, leaving deep physical and emotional scars on the community.[4]

Historical records show that numerous high-explosive bombs and incendiary devices fell across the area, resulting in widespread destruction. Thousands of homes in the borough were so severely damaged that demolition was required, profoundly reshaping the landscape and displacing thousands of residents. The human toll was equally severe, and hundreds of civilians were killed in Bethnal Green alone throughout the war, with many more injured or traumatized by the relentless raids.[5]

Yet from these scenes of destruction emerged remarkable resilience and renewal. The vacant bomb sites became accidental sanctuaries, gradually reclaimed by nature. Over decades, these urban scars transformed into vibrant habitats, demonstrating nature's tenacity and capacity for renewal even within densely human-populated areas. Among these revitalized spaces is the Bethnal Green Nature Reserve, once a bomb crater, now a living demonstration of how biodiversity can flourish in the most unexpected urban corners. The reserve encapsulates a hopeful narrative: humans and nature coexisting, thriving side by side, shaped by their shared history and mutual resilience.

In the post-war years, the scars of conflict remained visible. Bomb sites dotted the landscape, with some left derelict and others repurposed. One such site, on Middleton Street in Bethnal Green, lay dormant for decades until a group of local residents and artists saw potential in the rubble. They envisioned a space where nature could reclaim its place, and the community could find solace and connection.

This vision gave birth to the Bethnal Green Nature Reserve, a community-managed green space that transformed a former bomb site into a sanctuary of biodiversity and healing. Among those instrumental in this transformation was Michael Smythe, a London-based artist, curator and community gardener whose work explores the intersections of ecology, public space and social justice. In 2014, Michael co-founded Phytology, an urban physic garden and cultural institute based within the Reserve that explores the medicinal and social uses of plants in city environments. This initiative transformed a former Second World War bomb site into a vibrant community space, cultivating over thirty species of medicinal plants, many native to London and deeply rooted in local folklore, and fostering collaborations between artists, scientists and local residents. Through Phytology, Michael has created a living laboratory that challenges conventional notions of land use, highlighting the potential of urban green spaces to serve as sites of healing, learning and resistance.

Michael described the reserve as 'small, but tenacious and resilient,' mirroring the spirit of the local community, and emphasized the importance of forming relationships with the land: 'Once you start connecting with a piece of land, you start seeing how it changes over time, you start building a closer relationship with it. I feel like it can help you address a lot of things that feel impossible to even begin to take on, such as climate change.'[6]

Walk through the garden today and you'll find ribwort plantain (*Plantago lanceolata*). This plant, so ordinary you might step over it on a city pavement, is known for its antimicrobial properties and was once used to dress wounds. Yarrow, another of these so-called weeds, was historically packed into battlefield gashes when no bandages could be found; its name, *Achillea millefolium*, recalls Achilles himself. These humble plants have always carried the power to heal.

As Michael points out, recognizing the value of these plants is a political act. What the city regards as weeds, he calls 'critical infrastructure ... like schools and hospitals'. In a landscape where green space is shrinking and wellness is sold as a luxury, gardens like this offer something radical: care that is free, communal and rooted in history. I was gifted a small pot of the garden's balm – 'Good Skin', made from yarrow, ribwort plantain, marshmallow and chamomile – and discovered, as it soothed my eczema, that the knowledge embedded in these plants can still meet us where we hurt.

The ethos of the garden has stayed with me. It reminded me of the Yoruba deity Ọ̀sanyìn, the *orisha* (divine spirit) of herbal medicine and healing. Among the Yoruba peoples of West Africa, Ọ̀sanyìn is shown with one eye, one arm and one leg – a figure who, at first glance, seems incomplete. But that very depiction is the point: it is a reminder that what we dismiss as imperfect or insignificant often holds the deepest knowledge. When tasked with clearing a field, Ọ̀sanyìn refused to cut down the weeds, knowing that every plant holds its own medicine. In the Bethnal Green Nature Reserve, that same understanding takes root: that even the most overlooked plants have their own value and power.

*

If ruderal plants are nature's first responders – arriving to heal, to stabilize, to begin again – then perhaps they are also its witnesses. They grow in the places we try to forget: cratered fields, bomb sites, neglected verges. They root themselves into the very substance of upheaval, taking in the minerals, memories and scars left behind. They are alive in ways that are a contrast to monuments carved from stone. They grow, shed and flower, in cycles that reverberate with the histories buried beneath them. When we choose to garden with and alongside these ruderal species we are perhaps intuitively listening to the stories they have been telling us all along.

It was with this knowledge still settling within me that in 2024 I received a glorious invitation, one that would take the questions I'd been exploring through allotments, soil and seeds and place them in a very different historical and political landscape.

Professors at the University of the Western Cape in South Africa wanted to start a food garden – a garden that would both nourish students and staff physically and also tend to the complex histories of the land it would grow from. There was, however, a complication: the university's head gardener. A deeply warm and generous man, Jade was also, understandably, protective of the award-winning landscape gardens he had helped shape over many years. He would be following our activity closely, making sure that we knew what we were doing, and that we wouldn't unintentionally damage what he and others had so carefully nurtured. During the process of getting the garden started, Jade began by looking over our shoulders to see what we were up to and very quickly got stuck in helping us. We had organized a seed swap in one of the lecture rooms and we also wanted to work with soil. Jade kindly obliged and brought in sacks that we proceeded to tear open, releasing soil all over the floor and into which we could get our hands (and feet) so that we could begin to get our seeds going.

I had long been drawn to South Africa's history. At school in Kenya, we had created plays about 'the struggle', as it's known locally: the long and bitter fight to end apartheid.[7] Most people understand that apartheid was unjust, but it's in the details, the cruel bureaucracy of it, that the full horror becomes clear. During the struggle, the state would deliberately arrest Winnie Mandela just as her children were due to return home from boarding school in Eswatini, making sure that she couldn't see them. In detention, she was tortured repeatedly. And beyond the obvious brutality were the everyday, grinding violences: the passes, the exclusions, the constant reinforcement of a system designed to crush ordinary people's ability to live decent lives.[8]

As a way of bearing witness and standing in solidarity, our teachers taught us about key freedom fighters – Nelson Mandela, Steve Biko, Walter

Sisulu, Oliver Tambo – and we sang songs of resistance such as 'Nkosi Sikelel' iAfrika'. At Greenacres, a boarding school I attended in Limuru, it was even more personal. One of our favourite teachers, Mrs Rosemary Jacobs had escaped apartheid, moving to Kenya via Zambia. While we were trying to keep knowledge of the struggle alive in a pre-internet era, she had lived it.

I was 14 when Nelson Mandela was released from prison. Friends of my mother who had lived in exile for years were finally returning home, and they asked if we wanted to travel with them. We said yes, and the experience has left an indelible mark. At the time, de Klerk was still president, and the first democratic elections had yet to take place. No one knew whether the country would collapse into chaos. Everywhere we went, it was immediately obvious that we ourselves – Black people moving freely in areas where they had previously been banned – were not South African. The air was thick, the way *uji* (traditional East African porridge) thickens in the pot when you stop stirring.

When it was time to leave, an elderly man who worked at the B&B where we'd stayed ran after us, stopping our taxi. We thought we must have forgotten something. Instead, he pressed a postcard into my hand.

'I've never met another African from the continent before,' he said quietly, 'and I don't know if I ever will.'

It was a small, ordinary card – one of those free ones from a tourist rack – but the gesture was anything but casual. In that moment I understood it for what it was: a parting gift, the kind of offering you give when you have been moved by an encounter and know you may never meet again. That, too, is something deeply African: to send someone on their way carrying a piece of you, however small.

The chance to return decades later, this time to engage with the land as part of my professional life, wasn't something I could pass up. I worked closely on the garden project with three brilliant collaborators: Rory, whose heritage spans three continents – Africa, Asia and Europe; Mona, who is Middle Eastern diaspora, based in Southern Africa; and Robyn. Rory, a tall and relaxed professor with a passion for food histories, is one of the most attentive listeners I've ever met. On our daily drives, I noticed how easily he'd pick up a stray comment I'd made the day before and weave it deftly back into our conversation. Robyn, just as thoughtful, had an easy, infectious laugh that often carried us through heavier discussions. Mona was a bundle of determined energy, capable of holding complex ideas without losing clarity or heart.

Even though I'd read extensively before arriving, no amount of research could compare to standing on the actual grounds of the University of the Western Cape (UWC), on the windswept plains of the Cape Flats, a landscape shaped by centuries of upheaval and renewal. Long ago, these plains were thick with the biodiverse vegetation of coastal fynbos and strandveld. But colonial settlement took its toll. A region that had once included wetlands became canalized or infilled for residential settlement, leaving deep sand that was difficult to navigate. In the 19th century, colonial government officials planted Australian acacias to 'bind' the dunes – fast-growing, invasive trees that outcompeted native species. Much of the indigenous vegetation vanished beneath farmland and suburban sprawl. What had been one of the region's most widespread ecosystems now survives only in fragments, hanging on against relentless development.[9]

The human ecology was likewise upended. Apartheid planners, wielding the Group Areas Act, turned this area into a dumping ground for those cast out of 'white' Cape Town – entire communities were uprooted and forced onto the Flats. In the 1970s, informal settlements sprang up on the edges of UWC, only to be met with brutal forced removals: from January to August 1977, some 25,000 people were evicted from nearby Unibel and Modderdam squatter camps in bulldozer raids by the state. UWC itself, founded as a segregated college specifically for 'Coloureds',[10] soon defied its apartheid design – students and staff joined the freedom struggle, even aiding evicted families in camps like Modderdam and forging solidarity across racial lines. As Rory once put it, UWC was designed to train 'the invisible workers': nurses, clerks, accountants – the kinds of roles every society depends on, but which are too often undervalued or rendered invisible. Disciplines deemed intellectually or politically 'inappropriate' for Coloured people, such as archaeology, were simply excluded. To this day, UWC still has no archaeology department.

By the 1980s, UWC had proclaimed itself an 'intellectual home of the left', aligning academia with social justice and land rights activism. This history lives in the soil. A full quarter of the campus is now a nature reserve, established in 1977 to protect the dwindling strandveld and sand fynbos – refuge for hundreds of native plant species. The new food garden we helped to create carries this legacy forward, rooting new growth in old resistance.

*

The evening after the seed swap, I laid all the different varieties out on the table in my room. Seeds are such quiet things, self-contained packets of resilience and hope. They wait for the right conditions to arrive, even if it takes years.

The dahlia seeds caught my attention, partly because of their size and partly because they reminded me of sweet potatoes, which I love to eat. A dahlia tuber is an underground, starchy storage organ, brown and knobbly. From the crown at the top, tiny buds called 'eyes' form: the points from which new growth will sprout year after year. It's a plant built for survival, storing what it needs below ground until the conditions above allow it to rise again. They were brought by Azzara Makan, who runs heritage tours at the Jaftha family's flower farm in Constantia in Cape Town.

As I held those seeds, I kept thinking about the history they carried.

Dahlias are native to the high valleys of Mexico and Guatemala, where the Aztecs once grew them for food, medicine and for their hollow stems, used to carry water. After the Spanish conquest, they crossed the Atlantic, and when the first shipment of tubers arrived in Holland in the late 18th century, only a single one survived. From that lone survivor, an extraordinary profusion of new varieties emerged. That trajectory, from dislocation and near loss to abundance, has made the dahlia a living emblem of endurance and adaptation. Those seeds would not leave my mind, and so when Azzara offered to take us to see where they came from, I went.

Jaftha's Flower Farm lies in the Constantia Valley, on land from which the Jaftha family, like hundreds of other Coloured families, were forcibly removed in the 1960s under the apartheid Group Areas Act. For decades they were scattered across the Cape Flats – Parkwood, Retreat, Manenberg – while the farmstead and its palm tree stood empty. In the 1980s, Moses Jaftha, father of Charles and Malcolm, managed to return to Constantia, at first working as a butcher. With the help of a sympathetic farm stall owner, he leased a piece of land from the city and began again, planting flowers from the bulbs and seeds the family had managed to keep alive all those years. In 2008, they opened a small flower shop.

Walking through the fields today, you see row after row of blooms: dahlias, arum lilies, freesias, *Watsonia*, snapdragons. This farm is their pride, and the dahlias are the centre of it all, grown from tubers that have been with the family for more than 60 years, passed down like heirlooms. Charles and Malcolm learned everything by watching their parents and grandparents. 'The old people used to say: you steal with your eyes,' they told the group I was visiting with. And so they did, farming now as they did then, by hand, without pesticides, in continuity with those who came before.

As we walked, Malcolm told me a story that has stayed with me. During the long years after their removal, the family had buried some of their dahlia tubers in the very soil from which they had been driven, returning to dig them up and replant them once they were able to come back. Those flowers, blooming again in the same ground, are a living archive: proof that something can be taken away and still return.

The visit made me think of the Nubian families I have worked with in Kibera, who, like the Jafthas, found in the act of cultivating land a way to recover parts of themselves that dislocation had tried to erase. Looking at those dahlias, I saw how plants carry these histories, often silently, waiting – like seeds – for the right conditions to speak again.

As I read about and listened to stories in Cape Town about the gardens of the past, present and, crucially, the ones that would be growing in the future, I realized that the land itself was supporting recovery through an act of remembering. Each planting, each shared harvest, felt like a response not only to hunger, but to erasure. And it was through this deepening awareness, as I worked alongside others on the Cape Town garden, that an idea I had long circled began to crystallize – that plants are not just survivors of history, they are its witnesses. One person in particular provided me with the understanding *of* and language *for* this revelation.

The anthropologist Dr William Ellis had written about trees as witnesses: long-lived beings rooted in landscapes altered, sometimes violently, by generations of human activity. In his view, they were guardians of cultural memory, standing through upheaval, quietly recording what had passed.[11] His words helped me articulate something I had sensed but hadn't fully named in my own work: that plants don't just grow in a place – they register it. They absorb our histories. They hold the record of what we do, and what we fail to do.

That concept echoed powerfully again when I read about an ancient kauri tree unearthed in New Zealand, perfectly preserved for more than 42,000 years. Its rings offered more than just a timeline, they held evidence of a global climatic event: the brief collapse of earth's magnetic field. Scientists named it the 'Adams Event' – an extraordinary moment that may have changed the course of human history. What struck me was not just the science, but the poetry: a tree surviving through cataclysm, silently recording the world as it shifted around it.[12]

Thinking of plants in this way – as record keepers – brought new clarity to an old grief. As a child, I'd been heartbroken whenever a tree in our compound was felled. My mother, a pragmatic gardener who has planted thousands of trees, rarely hesitated when one needed removing. But I mourned them. One tree in particular stood just outside my bedroom window. I called it my time tree, not because it tracked the hours, but because its shadows marked the rhythms of my day. It stood watch over my secret garden, near the yesterday, today and tomorrow bush. When it was cut down, I cried for days. I hadn't lost just a tree. I had lost a witness – a companion who had seen my small daily rituals and anchored my childhood sense of time. It was only later, reading Ellis's work, that I understood why that loss had run so deep. The tree had been more than just a background to me – it had held memory. And not mine alone, but as part of a larger, shared remembering that lived in the landscape itself. In cutting it down, something more than physical had been severed. It seems that I was like my father in that respect. Once, in his work as a civil engineer, he even designed a building around a tree rather than cut it down. The hotel he was building stood on the Kenyan coast, a landscape where baobabs rise from the scrub like something both ancient and improbable. These giant trees, with their swollen trunks that can hold thousands of litres of water, their sparse crowns of branches like roots turned upside down, are some of the oldest living things in that dry coastal belt. For centuries they have been used as landmarks and gathering places, and in many cultures along the Swahili coast they are symbols of endurance: slow-growing, drought-resistant and able to regenerate after damage. One of those baobabs stood exactly where the hotel's reception had been planned. It could easily have been felled. Instead, my father worked with the architects to shift the plans so that the reception would be partially open, built around the tree, making it the centrepiece of the entrance. That baobab still stands there, holding the space he imagined. I still go back to visit it when I want to remember him.

*

Once I began to see plants as witnesses, I started to notice something else: they don't just stand by and watch. They become part of the work of recovery. Perhaps that is why so many people create remembrance gardens. A memorial made of stone asks you to come and look: a garden asks you to take part – to dig, to plant, to nurture. In doing so, it draws you into a relationship with something beyond yourself. Plants make that relationship possible. They grow alongside our grief, rooting memory into the ground and offering new growth in its place. Sometimes resilience starts this way, almost without us realizing it: a hand in the soil when there is nothing else to hold onto. Planting becomes a way to carry memory forward, not just as tribute but as a first step towards healing. I saw this powerfully in North Kensington in London, after the Grenfell Tower fire.

*

Tayshan Hayden-Smith didn't come to gardening through inheritance or horticultural training. He came through fire. When the Grenfell Tower blaze unfolded in 2017 in West London, Tayshan was abroad chasing a career as a professional footballer.[13] But that night fractured more than a building. He returned home to North Kensington and found his neighbourhood in mourning, the air still thick with ash and disbelief. In the days that followed, amid the shock and sorrow, an urgent question began to grow in him: what now?

The answer came, unexpectedly, through soil. Beneath the looming concrete of the Westway flyover, on a forgotten strip of land known as

Maxilla, Tayshan and others from the community began gathering. The space was neglected, the kind of urban scar tissue left behind by city planning indifferent to the lives around it. But in this no-place, they found potential. Collecting leftover plants from garden centres, coaxing colour from compacted dirt, they began to build what would become the Grenfell Garden of Peace. It was not just about flowers. It was about grief given form, and about hope insisting on growth.

Gardening, for Tayshan, became both sanctuary and strategy. Where trauma had scorched certainty, the act of planting gave rhythm and ritual. Each bed of herbs, each salvaged seedling, was a small movement towards an active form of healing. In the garden, the community gathered to remember, to share, to breathe. Soil became a language through which pain could be spoken, and beauty returned, tenderly and defiantly, to a landscape shaped by loss.

What began as guerrilla gardening has since grown into something enduring. Tayshan founded Grow2Know, a not-for-profit that champions inclusive green spaces and challenges ideas about who gets to belong in them. From West London estates to the Chelsea Flower Show, his designs refuse the quiet segregation of traditional British garden culture. They bloom with intention. At the heart of his work is the belief that green spaces should reflect the communities they serve, and that justice can take root in flower beds just as surely as it does in courtrooms.

Today, under that same stretch of motorway, Hope Gardens continues the work begun in the ashes of Grenfell. It is a place of remembrance, of resistance and of radical care. There is a green heart sculpture there, bearing the word 'Justice' – both a demand and an offering to the future. What Tayshan and his community created is a living monument. One that reminds us that tending to the land can also mean tending to each other.

*

Nature has always known what it means to adapt to disruption. Plants – especially the overlooked ones, the so-called weeds, the ruderal species – don't just survive in damaged landscapes; they begin the work of recovery before we even realize it has begun. They are nature's first responders. Their roots hold together disturbed soil. Their presence softens the sharp edges of crisis. Their blooms remind us that something beautiful can still emerge from what's broken.

And they do something else too: they watch. They record. They witness patiently without judgement. Trees remember through their rings. Mosses mark time in millimetres. Even the smallest seed carries within it a trace of the soil it once knew. In this way, plants are the surest form of archive, one that is a living presence to the nature around it, rather than cut off in a separate building. They know what has happened here, and they adapt in response. That, to me, is one of the most profound forms of resilience: to hold the past without being held back by it.

When we choose to plant, to tend, to dig – even in a pot on a windowsill or a patch of overlooked land – we are not just growing food or flowers. We are entering into a relationship with this more-than-human capacity for endurance. We are allowing nature to teach us something we have perhaps forgotten: that resilience rarely roars, it roots itself unobtrusively. It adapts. It holds on. And in doing so, it makes space for life to continue.

Working with plants, especially in spaces scarred by crisis – whether they are bomb sites, post-pandemic allotments, or neglected corners of the city – has shown me that healing is rarely linear. It spirals. It returns. It takes detours through memory and touch and sometimes grief. But always, there is growth. Often, that growth begins long before we see it. Beneath the surface, connections are forming, roots are reaching out, seeds are waiting for the right moment.

Joining with plants in this process – co-creating gardens, restoring damaged ground, or simply noticing what is already thriving in the margins – can feel like a small act. But it is also a defiant one. It includes us as part of this living, witnessing world. In doing so, we discover that resilience is something we can grow into, together.

Plants show us that survival is not the end of the story. What survives can thrive, not in spite of what it's endured, but because of it. And just as plants teach us to begin again from disruption, these gardens – and the people who tend them – remind us that healing and resistance can take root in even the most hostile terrain.

Together, these stories complicate the idea that nature and urban life are opposed forces. Instead, they reveal how cultivation, both literal and metaphorical, is a method of remembering, recovery and resilience. This is what it means to be human in the Anthropocene: not apart from nature, but entangled with it. These gardeners are more than simply stewards of the soil, they are cultural workers, memory keepers and visionaries – coaxing life, possibility and belonging from the overlooked cracks in the city.

Their garden spaces are not gestures of nostalgia. They are blueprints. Evidence that cities, even in their brokenness, carry the potential for regeneration – not in spite of human presence, but because of it. At the heart of each example is a reminder: tending to land is also a way of tending to memory, resisting erasure and building futures that are more just and more alive. What they suggest – and what this book will continue to explore – is that the story of biodiversity loss is not a foregone conclusion. Nor is it confined to distant wildernesses. If we look more closely, we see that biodiversity is also taking root in the places we've tended to ignore: in the city itself. And humans are not always the problem. Sometimes, we are the cultivators. Sometimes, the co-conspirators. Sometimes, we are the ones kneeling in public, pressing a seed into difficult soil, and trusting – not just that it might grow, but that it might help someone else breathe.

Chapter Seven

Beyond Rewilding: How City Insects Teach Us to Sustain Life

Alwin is lying on the treatment bench, and I am working through the tight muscles of his back when he says,

'I saw the leopard again. In Ngong Forest this time.'

This is what I love most about being an osteopath (before my career change to academia): there are always two conversations happening at once. One is the reason he's here – the stiff shoulder, the painful neck. The other happens while I'm treating. I talk to keep the body moving even when the work is uncomfortable, but these conversations do more than distract. They open a window into the patient's life. People tell you things without realizing, and suddenly you understand that it's the way they sit at their computer or don't move for hours when they're painting that is the real root of their pain.

'How close?' I ask, leaning into the muscles across his upper back.

'Close enough to see him watching me,' he says. 'There are at least two in the forest now. You should come.'

He works as an ecologist, so our conversations often drift this way – to animals, to the patterns of their lives across Nairobi.

He is always watching.

I tell him, 'I'd love to, but these days my free time is all toddler-time. A leopard walk and a two-year-old don't really mix.'

I move his shoulder through a slow stretch, then ask, 'Tell me about the bees again. I keep thinking of keeping some, but what do you do when you live in the middle of a school?'

He laughs. 'You don't.'

We talk about the woman in the news whose hives kept getting raided until she brought them inside her home.

'She and the children just got used to them,' I say.

'Bees adapt,' Alwin says. 'I've met beekeepers who have a way of working with them so they don't get stung. It's a kind of communication. I'm still trying to make sense of it.'

I press gently along the muscles either side of his spine and think of the school I live in. I can already picture the teachers' faces if I tried to persuade them that I could become a bee whisperer.

'It's what I miss most when I'm in the UK,' I tell him. 'Not just the big animals. Even the insects. My husband laughs at me because I still store food as though ants or weevils might get into it.'

'Habit,' Alwin says.

'But in the UK it can feel ... empty somehow,' I say. 'Why are there so few insects in Europe?'

He doesn't pause. 'Because we killed them all.'

He says it as though he were describing the weather. I feel the words land, heavy. For a moment, I lose my parallel thread. My hands keep working, but the conversation falls away.

*

Scientists across Europe have been monitoring the quiet disappearance of insects. They map what has been termed 'the windshield phenomenon', where drivers notice fewer insects splattered on their vehicles compared to previous decades. While the method makes sense, I struggle with the idea that we need to count the dead in order to know what might be living. A 20-year study in Denmark found an 80 per cent decline in insects collected on car windshields between 1997 and 2017, even after accounting for variables like time of day, date, temperature and wind speed.[1] Similarly, in the UK, the Kent Wildlife Trust and Buglife's 'Bugs Matter' survey reported a 63 per cent reduction in insect splats on vehicle number plates since 2021 and then again by by 62.5 per cent between 2021 and 2024.[2]

There's something deeply poignant in that: a smaller death toll on glass speaking to a much larger one in the world. The decline has been steep in recent decades, but it didn't begin there. It is the result of a long, cumulative process, one shaped, most of all, by the way we farm.

Across the European continent, landscapes have been simplified and sterilized in the name of production. Once vibrant field edges, full of thistles, wild carrots and hoverflies, have given way to monocultures stretching to the horizon. Pesticides are applied at industrial scale, to kill pests and often prophylactically, as part of a chemical routine. In Germany, a study found a 76 per cent decline in flying insect biomass in protected areas over just 27 years.[3] In the UK, long-term monitoring of farmland has shown dramatic drops in beneficial insects: ground beetles, ladybirds, lacewings – all victims of the same regime designed to make fields inhospitable to anything not profitable.[4]

Pesticides do not stop at the field's edge. They drift. They linger in the soil, accumulate in hedgerows and leach into waterways. And though their immediate targets might be aphids or caterpillars, they affect entire food webs, including the insects we rely on for pollination and soil health. France's ecological agencies have described the 'massive use of insecticides' as a primary driver of biodiversity loss.⁵ It's an understated violence, made palatable by its distance from the consumer.

This, then, is not a matter of nostalgia for a golden past, but a recognition of what is being lost – and the systems behind that loss. The more we push for maximum yield, the fewer insects remain to do the unheralded work of holding life together. And with their disappearance goes something of our own future.

I thought back to that conversation with Alwin years later, while working on my 1918 Allotment project. It was the only 'cheat' we allowed ourselves in recreating a wartime-era plot: we chose not to use pesticides.* At first, it seemed a simple, even obvious, decision. But the deeper I went into the history, the more tangled it became. When I researched what was used to deal with 'bugs' in 1918, I found myself tracing a grim line backwards – from modern insecticides to their early cousins: chemical weapons. Some of the first pesticides were rebranded wartime poisons, born from the same stockpiles of gas used on human beings during the First World War. French soldiers deployed tear gas to drive German troops from trenches; within years, variations of those same compounds were being sprayed on cabbage whites and caterpillars – even people. I've been tear-gassed myself, and I know what it feels like when the air turns against you, when breathing burns, your eyes sting shut and your body reels in protest. That experience changed my relationship with chemicals forever. When I see the delicate iridescence of a hoverfly's wing or the soft ticking legs of a ladybird on a leaf, I can't help but imagine what it means to be caught in a drift of poison, to have no mask, no shelter, no choice.

There's something profoundly unsettling in the way violence crosses categories. Historians like Edmund P Russell have argued that the metaphors and machinery of war don't end on the battlefield; they

---

* I reflected on this choice in an essay in my book *Portal: 1918 Allotment*.

bleed into the soil, the garden, the domestic.[6] Insects became enemies not through science alone, but through a way of thinking that framed nature as something to be subdued. When I practise permaculture growing, and plant sacrificial crops – plants I expect and even invite insects to eat – I realize I am afforded a different possibility. I have the privilege to wonder what collaboration might look like, to imagine a different relationship between humans and the many-legged lives that share our cities.

And yet awareness does not make the loss easier to overcome. The decline in insect populations across Europe has not happened overnight. It has unfolded gradually, almost imperceptibly, the result of decades of intensive agriculture, pesticide use and habitat loss, and a creeping disconnection from the living systems that surround us. Insects have vanished not with a bang but with a fading – a quiet absence where once there was movement, texture, sound.

This is the landscape into which ideas of 'rewilding' now emerge, often spoken of in hopeful tones, as a way to bring back life, to restore what has been harmed. At its core, rewilding is the idea of travelling back through time to return nature to a state before it was cultivated. But I find myself uneasy with how the term is used. Too often it suggests a return to an imagined past: a prehuman, or at least pre-industrial Eden, as though we might simply rewind history and find harmony waiting.

One of the most revealing, and sobering, examples is the case of Oostvaardersplassen in the Netherlands. Created in the 1960s and 1970s on land reclaimed from the sea, it was initially a wetland, but soon became a kind of ecological experiment – a flagship for rewilding in Europe. The idea was elegant on paper: reintroduce large herbivores to mimic prehistoric grazing patterns and let nature take its course, with minimal human intervention. Konik horses, Heck cattle and red deer were released to roam freely, shaping the landscape through their movements and appetites.

Fences kept them in, predators were absent and the philosophy was clear: step back, and let wildness return.

For a time, it seemed to work. The animals thrived, and the open grasslands flourished. But then the problems began to surface, at first slowly, and then unmistakably. With no predators to keep populations in check and no corridors through which animals could migrate or disperse, the herbivore numbers exploded. Overgrazing intensified. What had been a varied mosaic of grassland, wetland and woodland began to collapse into something simpler and more barren. Trees and shrubs disappeared. Bird diversity declined. The balance they had hoped would emerge never came.

And then came the winters. Especially harsh ones, in 2005 and again in 2017-18, exposed the deep vulnerabilities in the model. With food scarce and no way out, thousands of animals starved where they stood. Some were culled to prevent further suffering, but the scale was staggering. In a single winter, more than 3,000 animals died – over 60 per cent of the population. The public, understandably, were horrified. Protesters gathered at the reserve's fences. Activists threw hay over the barriers in defiance of the non-intervention policy. Debates about animal suffering, ecological ethics and the limits of human detachment filled the airwaves.[7] People had expected to see wild beauty. Instead, they saw bodies.

What makes the Oostvaardersplassen so important, so unsettling, is that it lays bare a central contradiction in how rewilding is often imagined. The project was framed as letting nature run free. But from the beginning it was shaped by very human choices: the fencing, the species introductions, the refusal to intervene. Even non-intervention is a form of management. And when the public demanded change, the policy shifted. Populations were capped. Active culling resumed. Relocation programmes were introduced. In the end, it wasn't nature taking its course, it was people navigating the uneasy edges of control and withdrawal.

To me, this is where the romantic idea of rewilding begins to fracture.

Without predators, without migration corridors, without truly vast and connected landscapes, we're not returning to something ancient – we're simulating it within constraints of our own making. And when things go wrong, the ethos of 'hands off' makes it harder to act. The animals suffer, but so too does the promise of rewilding itself. Rather than restoring balance, we may instead create new forms of imbalance, driven by an unwillingness to admit that humans are already part of the equation.

There's a strain of nostalgia embedded in rewilding – a desire to return to a time when nature was 'purer', less touched by human hands. But history offers no fixed point of perfection. Humans and nature have always been entangled. To suggest otherwise is to erase that long co-evolution, and to risk replacing one kind of exclusion with another. Too often, the 'wild' becomes moralized – a shorthand for what is good, pure or worth saving – without recognizing that much of what we think of as untouched nature has long been shaped by human activity. When I was a child, and though I didn't have words for it, I knew that the distinction between wild and tame was blurred, and more so than those who talk about 'rewilding' often allow. As I ate breakfast in Nairobi, looking out onto the veranda, there were mornings when I would delight in spotting a mongoose dashing across the tiles – something supposedly wild, appearing in our otherwise domestic setting. I liked the idea of sharing breakfast with a mongoose. That is, until my mother gently pointed out that its presence probably meant that the fresh, warm eggs we expected to collect from the chicken coop would already be gone. Her words split my heart. How could I make peace with this? I didn't want to lose the eggs, but I didn't want to lose the mongoose either. The answer, in the end, had to be practical – a sturdier chicken coop. But the moment stayed with me. The crossover was undeniable: we were already sharing our world. We always had been.

The lesson, I think, is not that rewilding is doomed, but that it must be honest. That in many settings – especially small, enclosed, human-dominated ones – non-intervention is neither possible nor desirable. Care cannot always mean stepping away. Sometimes, care means staying involved – attentive, adaptive and deeply aware of the power we hold and the consequences of pretending we don't. What if, instead of trying to erase ourselves from nature, we asked how to live better *within* it? What if the real work of restoration is not about absence, but instead a thoughtful, layered, participatory presence?

*

It wasn't until I was doing my doctoral studies that I began to find a language to question the ideas of 'wild' and 'tame' while working with my doctoral supervisor, who had carried out extensive fieldwork in lowland South America. It was through her that I came to understand that even the Amazon – so often held up as the ultimate symbol of wilderness – is not a pristine forest but a co-created landscape, shaped by thousands of years of Indigenous management.[8] Rather than disrupting the ecosystem, they enhanced its richness and resilience. This fundamentally shifted my understanding. If the most iconic wilderness on earth is in fact a cultural landscape, then what do we really mean when we call something 'wild'? Have we ever truly been wild ourselves?

So I offer a different frame. Instead of seeking to return to something imagined, what if we start from where we are? What if rewilding is not about going backwards, but about reimagining how we live together now? That requires individual effort, yes, but also citywide vision – structural interventions that only governments and large organizations can enact. Insects are a vital part of this story. They are often overlooked, yet their survival, especially in cities, points to the possibility of coexistence.

In this chapter, I'll trace how some cities are supporting insects in ways that rural areas now fail to. I'll explore what it means to build spaces not just for people, but for pollinators – to recognize that a truly inclusive city is one that hums, flutters and crawls with life. And, as always, I'll return to the stories we tell about these creatures, from myth to modern science, to ask: what does it mean to share the city not only with each other, but with those whose buzz we often ignore, like that person who irritates us at first but who turns out to be essential – the kind we learn to love once we realize that we can't live without them?

*

For a few years, I lived on a street in Oxford that had a particularly active group of guerrilla gardeners. They would push leaflets through people's doors, inviting them to join work parties that planted wildflowers, lined front walls with hollyhocks, and tended to flower beds around the base of the street's trees. I'm indebted to them. Simply by observing their work over the years, they taught me how to *read* a space that had been guerrilla gardened – a skill that not only deepened my enjoyment of walking through the city but also became a critical tool in my doctoral studies.

But they didn't stop at planting. Over time, the guerrilla gardeners began to shape the street's ecology more deliberately, speaking directly with the council about how local nature should be managed. One spring, handmade signs appeared among the wildflowers circling the base of the trees, carefully lettered messages that read 'Bee Café' and 'Do Not Spray'. And, remarkably, the council listened. They stopped using herbicides on that street. The result was subtle but unmistakable: more blooms, more buzz, a softness underfoot that wasn't there before. Even now, when I walk down that stretch, it feels different from the surrounding streets – richer, more alive. Watching those neighbours advocate for the right of bees to exist in the city changed the way I thought about pollinators. These weren't just acts of quiet beautification, they were forms of activism. Their signs weren't just whimsical, they were arguments. Declarations, even. They asked us to see insects as legitimate residents of the city, worthy of protection. And in doing so, they reminded me just how easily I'd taken pollinators for granted, perhaps because I had spent my childhood in a city where their presence was never in doubt.

I also spoke with other people in the city who were actively making space for insects, like the allotmenteers, especially those who still bounded their plots with traditional flower borders or had set aside pollinator patches. Barbara, a grandmother and passionate grower, told me how much hers had come to mean to her over the years. Sitting in the shade of her shed with a cup of tea, she'd take elevenses 'with the bees', as she put it, watching them sip nectar while she sipped hers. 'It makes me feel connected to the wider world,' she said. And I knew exactly what she meant.

We often say that we can't live without pollinators, and that's both true and not entirely true. Most of the staple crops we rely on – wheat, rice, maize – don't require animal pollinators at all. They're wind-pollinated or self-pollinated. The same goes for many of the vegetables we eat for their leaves, stems or roots: spinach, lettuce, onions, potatoes, carrots. But where pollinators *do* matter, they matter profoundly. Apples, almonds, cherries, courgettes, cucumbers, pumpkins and a riot of soft fruits depend on bees and their kin for high-quality yields. Without them, our diets would be duller, less nutritious, and our ecosystems unbalanced. Perhaps more importantly, the presence of pollinators is a sign that a place is hospitable – not just to bees, but to life in general.

We've known about pollination for a long time. In ancient Assyrian art, for example, reliefs depict winged deities engaging in what appears to be the pollination of sacred trees, possibly date palms, using pine-cone-like objects. These images suggest that ancient cultures recognized the importance of pollination in agriculture and perhaps even in ritual contexts.[9] Fast-forward to the 18th century, when Christian Konrad Sprengel systematically studied the role of insects in pollination. In his 1793 work, *Das entdeckte Geheimnis der Natur im Bau und in der Befruchtung der Blumen*, Sprengel detailed how floral structures and colours attract insects, facilitating cross-pollination – a concept that was revolutionary at the time.[10] And I like to think that the First World War allotmenteers who bordered their plots with flowers – even when every square inch of soil was precious in preventing them from starving – knew what they were doing. Maybe they didn't leave diaries explaining their choices, but the flowers themselves tell a story. They remind us that beauty, function and cohabitation aren't mutually exclusive.

That's why Oxford's transformation into a haven for pollinators feels so meaningful. The city council changed its green space management strategy, letting verges and park lawns grow wild, replacing uniform turf with wildflowers. The shift is part of a deliberate Pollinator Action Plan, developed with groups like Wild Oxfordshire and the local Wildlife Trust. It's a lot more than a cosmetic change: it's a structural reimagining of what city spaces can do and who they are for. Once-neat strips of grass now host bees, moths, hoverflies and butterflies. Wildflower patches have replaced the aesthetic of order with something richer, messier and far more alive.[11]

This work extends beyond policy documents. Oxford University's Wytham Woods research team launched the Oxford Plan Bee project, which connects ancient woodland to school gardens through community planting. Led by Dr Kim Polgreen, the initiative involved students from local schools planting bee-friendly flowers in neglected parks and playgrounds. They created what's been called 'pocket rewilding', transforming small urban plots into nectar-rich oases. These efforts, supported by local councillors and government grants, reflect a layered approach: scientific knowledge, community engagement and institutional commitment all working together.[12]

And it's working. Bee populations in Oxford are doing better than those in many surrounding rural areas, where monocultures and pesticide use dominate. Surveys report a higher diversity of wild bee species in the city's parks and gardens, where flowers bloom across seasons and insecticides are seldom used. It turns out that in a time of biodiversity crisis, the city – when thoughtfully planned – can offer a refuge. And not just for humans, but for all the other lives we so often overlook. The very phrase 'pocket rewilding' suggests something different from traditional rewilding; something smaller, more intimate, and perhaps more radical in its attention. A pocket is close to the body. It holds what we need to keep safe, keep near. These pockets of green are not about restoring some imagined wilderness, but about stitching new relationships into the fabric of our daily lives. They carry a different kind of intention. 'Pocket rewilding' isn't about stepping back to let nature take over; it's about stepping closer, asking how we might live alongside the lives that persist and emerge in the gaps we notice enough to care about. It's not a return. It's a redesign – one rooted in reciprocity, proximity and care.

It's not only Oxford that is rethinking its bond with pollinators. Cities across the globe are beginning to attune themselves to the essential presence of bees, butterflies and other insect life. In Singapore, known evocatively as the 'City in a Garden', this reimagining of what a city can be has been carefully and deliberately cultivated. Here, urban biodiversity is not an afterthought but a guiding philosophy. There has been an evolving awareness that supporting nature need not be at the cost of progress. Today, green arteries weave through the dense urban fabric, living corridors that nourish both people and pollinators. The dazzling Gardens by the Bay, a marvel of renewal built atop reclaimed industrial land, blossoms with life at the very heart of the city. Even concrete channels have been softened into living waterways, like the stream now running through Bishan-Ang Mo Kio Park, where biodiversity soared by 30 per cent in just two years – without the need to reintroduce a single species.[13] Life, it seems, was waiting.

Singapore's commitment is both poetic and strategic, layered and deeply intentional. Its 'Nature Ways' span over 90 miles, tree-lined corridors that thread the city's parks and reserves together like green stitches on a dense quilt of urban life. These routes are more than scenic walkways; they are lifelines for butterflies and bees, carefully planted with native flowering trees and shrubs to act as aerial passageways through the cityscape. The National Parks Board, or NParks, leads the charge, blending ecological science with community engagement. Through their programmes, citizens are encouraged to plant gardens that sing in the frequencies of pollinators, offering nectar-rich flowers and nesting sites. Singapore is now home to more than a hundred native bee species, from industrious honeybees to the almost imperceptibly small stingless varieties. With initiatives like 'Befriending the Bees', NParks reminds us that these miniature citizens belong just as much to the city as we do – and that our future is intertwined with their calm, ceaseless flight.

The outcomes are evident in Singapore's urban ecology. Despite its highly urbanized environment, 133 species of bees have been recorded in Singapore – a surprisingly high diversity.[14] Observations in city parks have found many wild bee and wasp species, including new records for the country. Singapore's holistic approach, sometimes called the City Biodiversity Index or 'Singapore Index' when exported to other cities, means that even as skyscrapers rise, pockets of pollinator habitat form part of the tapestry. The government drives these efforts through top-down planning and funding, but they also encourage community stewardship. For example, community gardens are supported throughout Singapore, giving residents space to grow flowering plants and learn about pollinators. Through this layered approach, Singapore has earned a reputation as one of Asia's greenest cities – proof that dense urban development can still sustain rich insect life. It's a kind of urban ecological attunement, where planting, policy and care form an organizing principle to create the conditions for life.

It's something I haven't been able to shake: the possibility that nature doesn't disappear but lies dormant, patient, listening. I once watched a video – shared with a certain breathless alarm in a WhatsApp group – tracing Singapore's birth from rainforest to metropolis. Bulldozers carving the land, steel and concrete pouring into what was once tangled jungle. And yet, beneath that dramatic transformation, what if the seeds of nature remained, scattered and unseen?

That same thought arises each time I peer into our garden water features at home in Oxford – two halves of what had been a whiskey barrel, upturned and filled with water and aquatic plants. By midsummer the frogs came. We hadn't introduced them. We hadn't done anything, really, except offer a welcome. And they responded. As though the invitation alone was enough. What would our cities become if we all dared to issue such invitations? Small gestures – planting, watering, pausing – each a quiet signal that other forms of life are more than just tolerated, they are also wanted.

*

All over the world – in cities as different as Dublin is to New York, or Nottingham is to Harbin in China – there are stories of success under the broad and often loosely applied banner of 'rewilding'. While I appreciate what this can mean for ecosystems, I find myself, as I've mentioned, agreeing with Kevin Sloan when he points out that 'the fiction advanced by re-wilding is the memory of an intended landscape type – a woodland, wetland, prairie – that is heightened by suppressing the appearance and evidence of human hands.'[15] Rewilding is, after all, still a human interaction with nature. What I find problematic is not the practice itself, but the notion that removing human presence or influence somehow makes a space more authentic. This creates yet another false divide between people and nature – an idea that nature is purer, better off, when we're not in it. That's a dangerous story to tell.

As an anthropologist of the imagination, I'm drawn instead to what historian Simon Schama describes in *Landscape and Memory*: the idea that we read meaning into natural forms based on our experiences, and the collective and individual memories we inherit.[16] These interpretations are passed down through generations. The way I see it, when we read a landscape, we are reading the dreams and thoughts of our ancestors – although they are written in very particular languages. It may be, for example, easier for someone British to read the manicured lawn of a North American suburb as a faint memory of an English estate, than to read a patch of the Amazon rainforest as a cultivated fruit garden planted by the ancestors of the people still living there.

*

It was another allotmenteer who first helped me think more deeply about why cities might actually be good homes for insects, beyond my own sense that humans and nature have always been intertwined. Paul was a tenacious grower. He tended a plot on a site that had never been properly fenced; it began informally and the council never quite took responsibility for it. Three years in a row, his entire harvest was stolen. And yet, despite that, he didn't give up. His plot was thriving, even if someone else was enjoying the fruits of his labour. I spent a day shadowing him at the nursery where he worked in Oxford, watching how his hands moved instinctively among the plants, how he carried decades of practical knowledge in the way he pruned, watered, paused. I told him how much I loved both cities and nature, and how I couldn't separate the two in my mind. He smiled and said something that has stayed with me: 'Plants love cities too – you just have to notice. Buildings give them shelter. They get a bit of warmth, a microclimate.'

It made perfect sense, though I hadn't seen it that way before. His words came back to me when I visited the Deutsches Museum von Meisterwerken der Naturwissenschaft und Technik (the German Museum of Masterpieces of Science and Technology). There, I learned that reinforced concrete – concrete strengthened with embedded materials like steel – was refined into the form we know today by Joseph Monier, a French gardener. Monier had started embedding iron mesh into concrete to make sturdier garden tubs and planters. He patented the idea in 1867 and eventually expanded it to bridges, beams and floors.

So reinforced concrete – often and understandably maligned – was first developed by someone simply trying to make the perfect flowerpot. That changed how I looked at cities. Cities are more than spaces to endure or escape from, they are vast containers for life. Giant pots, if you will. The real challenge, of course, lies in how we shape the container – and who we allow to flourish within it.

One of the more surprising findings from recent research is that bees, in many cases, are thriving in cities more than they are in the surrounding countryside. Oxford, where Paul and I tended our plots, is a case in point. Despite being surrounded by rural farmland, its urban environment supports healthier and more productive bee colonies than many nearby agricultural areas. Several factors help explain this urban edge.

Cities provide a continuous and diverse source of forage. In Oxford, a bee might pass crocuses in spring, lavender in front gardens in summer, and ivy blooming on old college walls well into autumn. In contrast, the fields outside town often offer a single floral burst – like oilseed rape – that blooms briefly, then disappears. A study in southern England found that urban bumblebee colonies had higher reproductive success than rural ones, in part because their diets were more varied and their nectar supply more consistent.[17] That botanical diversity, even from ornamental or non-native plants, leads to better bee health and stronger immunity.

Cities also tend to be lower in pesticide use. Oxford's council has reduced spraying in public green spaces, in part from the persuasiveness of guerrilla gardeners and in part to protect pollinators. Meanwhile, the countryside continues to suffer from the legacy of industrial farming, where chemical treatments remain common. For all their concrete and congestion, cities often offer bees a safer, more hospitable world.

There's also the benefit of warmth and shelter. Cities create their own microclimates, often a few degrees warmer than the surrounding land. This can help bees extend their foraging season, and it also means more nesting opportunities. Oxford's old stone walls, shed corners, compost heaps and even church towers become homes for solitary bees and wild swarms. In Oxford, the very fabric of the city – historic, messy, irregular – offers niches that the countryside, with its monocultures and heavily regulated farming landscapes, no longer does.

This all serves to flip some of our romantic assumptions on their head. We like to think of the countryside as inherently rich with life and the city as its enemy. But that binary is too simple. With the right choices – wildflower verges, chemical-free gardens, flowering balconies and bee-friendly parks – cities can become vibrant ecosystems. With enough care, our towns and suburbs could become 'giant nature reserves' for insects. And Oxford's bees are living proof.

*

In cities around the world, bees are welcomed like honoured guests – charmed, coddled and given boutique hotels made of bamboo and drilled timber. Wasps, meanwhile, are met with flinches, swats and a call to pest control. If a swarm of bees takes up residence in an inconvenient spot – an attic, a school compound, the eaves of a house – it is not unusual for a beekeeper to be summoned. The bees are gently gathered, re-hived and whisked away to safety, often with a small sense of ceremony. When wasps build a nest in the same place, their fate is usually sealed. The exterminator comes with smoke or foam, and the colony is wiped out. We treat bees as creatures to be saved; wasps as problems to be solved. This bias is so deeply ingrained that it barely registers. Bees are pollinators, food-bringers, fuzzy emblems of industrious gentleness. Wasps are just angry summer spoilers with a sting. But what if we've misunderstood the story? What if our aversion is simply habit formed by an old narrow narrative that no longer fits the world we live in?

Feelings about wasps, it seems, are near universal. Studies have shown that wasps are broadly and consistently disliked. When asked to describe them, people reach for words like 'angry', 'aggressive', 'dangerous'. Bees, by contrast, are praised as 'industrious', 'gentle', even 'cute'. You can spot bee motifs everywhere – in art, on clothing, on all sorts of decorations – but this is not true for wasps. One survey found bees rated more favourably than butterflies – an astonishing feat when you think about it – while wasps languished near the bottom of the likeability list. But the fact is that both bees and wasps can sting. The difference lies in the meaning we attach to this. Our dislike of wasps is shaped by the few species that cross into our lives: mostly the social wasps, yellowjackets and hornets, that build their papery nests under eaves and picnic tables. There are only about 67 species of these globally, a tiny sliver of the wasp family. But they are visible and loud, and so they shape our perception. The truth is that most wasps, more than 75,000 species, are solitary, and they are rarely to be seen, much less found stinging anyone. Yet we've let a handful of bold characters define the whole cast. Bees, by contrast, benefit from the opposite effect. The honeybee – just one species among over 20,000 – is the face of the entire bee world: fuzzy, golden, self-sacrificing (we like to comment on how it dies when it stings, as if to reassure ourselves that it doesn't really want to hurt us). We've poured meaning into the bee. Think of all the idioms: 'busy as a bee', 'the bee's knees', language soaked in admiration. Wasps, on the other hand, are cast as sly or cruel, stinging for sport. Even science is skewed by this perception: nearly 98 per cent of published research on stinging insects focuses on bees, while wasps receive only a tiny fraction of that attention. It's a bias that runs deep – emotional, cultural, even institutional.[18]

If I were being flippant, I'd say we could all learn a thing or two from bees about public relations. But it's probably more to do with what the psychologists Amos Tversky and Daniel Kahneman call the representativeness heuristic – our tendency to judge the whole by the part.[19] It's like visiting a country for the first time and forming your opinion of the entire nation based on the one taxi driver who picked you up at the airport. If they're warm and generous, you might walk away thinking the whole country is full of kind people. The insect world is so vast, so intricately diverse, that we inevitably draw conclusions from the few species that cross our path. And because most of us meet the bee with her golden fuzz and polite sting, and the wasp with his angular menace and picnic raids, we think we know who's friend and who's foe.

But despite this PR disparity, wasps are just as entangled in the life of the city as bees are – and just as essential. We already know that bees are vital pollinators, ferrying pollen between wildflowers and food crops, keeping community gardens blooming and urban fruit trees bearing. But wasps, for the most part, are hunters – garden guardians in disguise. A social wasp colony in a park or allotment might consume thousands of flies, caterpillars and spiders over a season, feeding their larvae with the insects that would otherwise munch their way through our lettuces and runner beans. That paper wasp cruising above your tomato patch? She's probably on the lookout for a juicy green caterpillar to carry home, not you. They're pest control on wings, paid in sugar rather than salary, our overlooked allies in the task of urban balance.

And while we don't tend to think of wasps as pollinators, the reality is more nuanced. True, they're not as efficient as bees: their smooth bodies don't pick up as much pollen, and they visit flowers more for a sugar rush than a pollen load. But some wasps *do* pollinate, and in certain cases they're the only ones who can. Tiny fig wasps crawl into the dark chambers of a fig's hidden flowers to complete their ancient compact. Some orchids have evolved to attract only wasps, tricking them with scent. And a whole family of wasps – Masaridae, or 'pollen wasps' – has taken things a step further, feeding their young pollen instead of meat, much like bees do. These blurred lines remind us that nature doesn't always colour within the lines we draw. In terms of ecosystems, bees and wasps aren't locked into separate roles, they overlap, and together they create a more resilient, more responsive urban ecology. A study in *Ecological Entomology* called for valuing both as natural capital: bees for pollination, wasps for biocontrol.[20] One feeds the fruit; the other protects it.

Yet in cities that nuance often gets lost. Bees are saved, resettled and serenaded; wasps are sprayed and smoked out. But that may be starting to change. Conservationists are beginning to argue that if we're serious about biodiversity, we can't just focus on the beautiful and benign. All insects are under pressure – from habitat loss, from pesticides, from the creeping shifts of climate. Urban biodiversity plans, like Oxford's Plan Bee, have started to name wasps alongside bumblebees, moths and flies, deliberately expanding the pollinator pantheon. And, as always, education plays its part. When people learn that the wasp at their sandwich may have eaten dozens of crop-damaging caterpillars that week, they sometimes reconsider. The ethnologist Dr Alessandro Cini puts it simply: we need a complete cultural shift to extend the kind of affection bees enjoy to their unloved cousins. But shifts are possible: if we tell better stories, if we share the facts, if we make space in our imaginations. After all, the city of the future won't just need bees. It will need the hunters and the gatherers, the fuzzy and the fierce, all working – often invisibly – by our side.

Maybe, by learning from our long mythological entanglement with bees, we can begin to imagine what kinds of stories we might need for wasps. Because our relationship with bees hasn't only been ecological – it has been symbolic, spiritual, deeply cultural. Bees have flown not just through meadows and orchards, but through our myths, our rituals, our ideas of what it means to live well and attentively in the world. Perhaps that's part of why we protect them. We have been telling stories about bees for thousands of years, stories that gave them a place in our inner lives.

In ancient Egypt, bees were said to be born from the tears of Ra, the sun god. As his tears fell and touched the earth, they transformed into bees – creatures of golden light, connecting heaven and soil. They were seen as symbols of resurrection and royalty, messengers between worlds. San peoples of the Kalahari have a myth in which a small bee carries the mantis god across a great flood, flying until it can go no further. With its last breath, the bee plants a seed in the mantis's body. That seed becomes the first human. A creation story, rather than highlighting dominion, instead focuses on sacrifice and cooperation. The smallest creature carrying the future of life within it.

In Zimbabwean folklore, the honeyguide bird once asked the bees for a bride. He had watched them closely – how they worked together, how their hive thrummed with sweetness and purpose – and he longed to be part of that world. But the bees turned him down, refusing to share their kin. Angered by the rejection, the honeyguide plotted revenge. He began to lead humans to the hidden hives, calling out with a distinct, beckoning song, fluttering just ahead to guide them through the bush. When the humans broke open the hive and took the honey, the honeyguide would feast on the wax and leftovers.

It's a story with layers – cleverness, bitterness, consequence – and it rests on something very real. In life, the honeyguide bird does just that: partners with humans to find wild hives, then waits patiently for its share. The tale encodes knowledge and practice into narrative, showing us how people once read the signs of the wild and worked with them, not apart from them, to survive.

And then there's the quiet tenderness of telling the bees – a centuries-old tradition primarily found in Western Europe, common in rural communities and still whispered about in some corners of the countryside today. Though its beginnings are hard to trace, the custom is frequently tied to Celtic mythology, where bees were believed to carry messages between this world and the next. The custom is most commonly documented in England, though variations have been observed in Ireland, Wales, France, Germany, the Netherlands, Switzerland, Bohemia (now Czechia) and even parts of the United States. When a death occurred in the household, it was customary to inform the bees, often by draping the hives in black cloth and softly speaking the news – sometimes addressing the bees by name, or tapping gently on the hive to get their attention. They were also told of marriages, births and even changes in household ownership. Failing to tell the bees was thought to bring misfortune: the hive might stop producing honey, fall ill or abandon its home entirely. The endurance of this practice well into modern times is striking – after the death of Queen Elizabeth II in 2022, the royal beekeeper at Buckingham Palace was reported to have formally informed the palace hives of her passing and of King Charles's accession. The ritual anthropomorphizes bees in the most intimate way, attributing to them awareness, loyalty, grief, and a role in the household's emotional life. Whatever its origins, the act speaks to a time when bees were seen as members of the household, silent witnesses to human joy and sorrow, and kin in their own right.

> *Small wings bear pollen*
> *life carried on beating air*
> *we begin as kin.*

What all these stories have in common is the way they draw bees into the human realm as co-travellers. They are given personality, purpose, even moral agency. And maybe that's what we haven't done for wasps. We haven't yet imagined them into our stories in the same way. They sting, yes – but so do bees. What they lack, perhaps, is a mythology of meaning, of reciprocity. No ancient god is carried to safety on a wasp's back. No one drapes a hive of them in mourning cloth.

But maybe we can change that. Maybe the next phase of urban ecology isn't just about reintroducing species – it's about reintroducing story. About learning to see all insects not as background noise or seasonal nuisance, but as participants in the life of the city. If myths once helped us live alongside bees, perhaps we now need new ones to help us live alongside wasps and understand them better. And also to learn from the way they work to bring about balance in our shared world. This is because a city isn't just steel and glass. It is a vast, living container, full of soft edges and wild corners, where unseen work is always being done. The wasps are already here, patrolling their paper nests, snipping at caterpillars, sipping sugar from fallen fruit. They ask for little. They offer more than we realize. Perhaps it's time we started telling their stories too because stories can help us love what we once feared and they are the groundwork of a future where we live together well. If rewilding asks us to remember, story asks us to imagine. And in cities that buzz, crawl, sting and shimmer with life, that act of imagination might be the most sustainable thing of all. But what happens when our gaze shifts downwards – away from the sky and into the damp, fertile ground where water lingers? The answer lies in the wetlands: those overlooked, life-bearing places on which civilizations have always depended.

Chapter Eight

Soft Civilizations: How Wetlands Shaped Urban Worlds

We're off to see grandfather in Kosano. Joseph did that cool big brother thing before we left Nairobi and was telling everyone that we were off to The Coast. No one takes him seriously because The Coast in Kenya means the amazing Swahili East Coast, with kilometres of sandy beaches and out-of-this world hotels. We originate from the west coast of Kenya. No one even calls it a coast, even though Nam Lolwe (Lake Victoria) is just a short flight away for a crow.

Right now it's the rainy season, which means one thing: chaos. The road has turned into a kind of mud river, and our family's old Peugeot 504 station wagon lurches and groans like it's trying to give up. The mud rises up to meet the wheels, sucking them in like something alive. Part of me still thinks it's brilliant, until I remember that I'm not small enough to hide in the boot any more. Now I have to get out and help push. I'm not going to complain. Everyone else is pushing and I don't want to be the spoiled Nairobi girl. But the thrill wears off fast. Mud in my shoes, hands slipping on the boot, the whole car shuddering as we try to shove it forward by force. Now it's not an adventure. It's just hard.

I think instead about the things that I'll look forward to when we finally get there. The delicious rice and vegetables that grandmother makes, drinking bright orange Fanta (usually banned by my mother) straight out of the glass bottle. The gorgeous black and white studio photos of all extended family members that cover the walls of grandfather's house that I can look at for hours. Great Auntie who makes the best roasted groundnuts and always says the quiet part out loud – thanks to her I get to find out exactly what it is the adults are really thinking and because of her age no one can tell her to shut up. For a moment it hits me that being in the village is like an in-between world. There are parts of life there that have not changed since my grandfather was a boy while other parts are no different to Nairobi. Suddenly, the car lurches forwards and we all have to catch ourselves to avoid face-planting into the mud. I can tell that Daddy is really happy to be home.

*

Home or 'dala', as we call it in Luo, is part family and part an ongoing relationship with a specific body of water called Aora Nalo (the River Nile). No matter where we might live, it is how we orient ourselves. I am a daughter of the Nile. My people migrated down the length of that great river over millennia, following its rhythms and riches until we arrived at Nam Lolwe.[1] When I think of wetlands, I don't think of an abstract ecological category, I think of home. And not in the metaphorical sense, but in the mud-on-your-shoes, dragonflies-in-the-air, fish-in-the-basket way that wetlands offer.

On the one hand wetlands provide things that I love, like the clay that can be fired to make beautiful pots, but wetlands near the Nile also host mosquitoes that can spread malaria. The gifts that wetlands host are not entirely cost-free. They are a constant reminder that nature is a complex mix of what we find helpful and what can harm us, and that how we choose to navigate this reality is what shapes our lives.

※

It turns out that scientific research has quite a lot to say about the benefits of wetlands. Wetlands are some of the most ecologically valuable landscapes on earth – rich, productive and astonishingly efficient. They cover just about 6 per cent of the planet's surface, and yet they store twice as much carbon as all the world's forests combined. Peatlands, mangroves, swamps and marshes act like giant green lungs and carbon sinks, helping regulate our climate in ways most of us are not aware of. What I used to think of as just 'swampy ground' is in fact one of the planet's most powerful defences against global warming.

They're also water purifiers. As floodwaters or rain run through wetlands, sediments settle and pollutants are absorbed or broken down by plants and microbes. It may not be glamorous, but it's miraculous. Without machinery or chemicals, wetlands clean the water that flows through them. They are natural filtration systems that cities spend millions trying to replicate in concrete. In fact, the very same mushy, mosquito-ridden landscapes that people once drained in the name of 'progress' are now being restored in parts of Europe to fix the damage caused by overengineering.

Then there's the matter of flooding itself. When rivers burst their banks or rainfall surges beyond the usual, wetlands act like massive sponges, soaking up excess water and releasing it slowly over time. They protect local communities from floods and help to stabilize entire river systems. In a time of climate crisis and erratic weather, we are beginning to realize that every hectare of wetland destroyed increases our vulnerability.

And, perhaps most profoundly, wetlands are cradles of life, and, as I will go on to argue here, of civilization. They host more biodiversity than almost any other kind of ecosystem, as home to amphibians, birds, fish, mammals and, of course, insects. Think of the millions of migratory birds that rely on wetlands for breeding or feeding grounds. Wetlands are not just stops along the way – they *are* the way, sustaining life in every direction. When I was younger, I didn't know any of this in scientific terms, but now I understand more deeply that the wetlands weren't just a backdrop to my childhood – they were doing the quiet work of holding the world together.

※

Ahero (the nearest town to Kosano) is full of rice paddies. It didn't occur to me as a teenager how strange it is that we often see food production and wetlands as separate or opposed. For Luo and many wetland peoples around the world, they are the same thing. Wetland soil is fertile, nutrient-rich. The cycles of planting, fishing and cattle grazing were once in sync with the seasonal moods of the water, but this process was profoundly disrupted by the historical period of colonialism.

I was an adult before I fully appreciated that we are, in every sense, wetlands people.

And yet wetlands are among the most maligned landscapes in the world. In much of Euro-American folklore, it's not just the landscape itself that is feared, it's what's imagined to live there, what might rise up from the depths. These places become the setting for menace, in the shape of a water hag with green skin and sharp teeth, for example, who lures children and the unwary into the abyss. Then there are the will-o'-the-wisps, those ghostly lights that flicker over marshes, said to lead travellers astray, deeper into the mire. And sometimes it's the dead themselves that rise – peat-preserved bodies surfacing centuries later, blurred by myth into restless spirits, reminders that these landscapes remember everything.

Why do these myths persist? Perhaps because wetlands refuse to be contained. They are neither land nor water, neither dry nor fully submerged. They shift, they flood, they absorb. They don't fit neatly into our maps or municipal boundaries. They resist the lines we draw, and that resistance unnerves us.

But that is exactly what I love about them.

In 2024, I was the inaugural artist-in-residence at Earth Trust in Oxfordshire. The landscape there is shaped by the Thames, which has undergone a kind of quiet revolution. Instead of being canalized and straightened into submission, the river has been allowed to meander, to move like a lazy serpent. One of the team told me, with delight, that it is now wibbly and wobbly. I felt that it looked like a series of interconnected Ws. Yes, wiggly, and in that wiggle lies its wandering wisdom.

When water is allowed to move - when it meanders, pools, spills - something ancient stirs. We think of straight lines and hard surfaces as the hallmarks of progress. But what if civilization, at its most enduring, has always been soft? Shaped by water, responsive rather than rigid. Wetland cultures don't just endure the wet. They understand that life is not a thing to be built in spite of water, but because of it.

I am biased, of course, but wetlands have been good to us Luo peoples. We are known - sometimes stereotyped - for our love of education and the good life. I like to think this has something to do with the abundance that wetlands provide: fish, fodder, fertile land. When you are not scraping a living from barren soil, you have time for music. We have dozens of traditional musical instruments. We have elaborate poetry, storytelling and theatre. The time we have stored up for that good life even seeps into our language. There are dozens of Luo words to describe exactly how the mouth interacts with food and drink. For example, *madho* means 'to drink' but *ruoyo* is 'to drink quickly', especially if it is a cold drink, while *uso* is 'to carefully sip a hot drink' - the list goes on. Being an anthropologist, I see in these cultural riches the traces of an economy that allowed for leisure, creativity and complexity. In this way, language becomes a kind of infrastructure too – a mental canal system, shaped for connection, retention and relationality. These aren't just poetic flourishes. They're tools for ecological survival. The words we inherit help us notice the world differently. They remind us how to live with water, how to dwell in flux without losing our footing.

Even now, far from Nam Lolwe, I feel the pull of wetlands. I live in the city, but I carry the lake inside me. When I see dragonflies dancing above the canal or the shimmer of reeds in a forgotten corner of the park, something ancestral stirs.

Wetlands are not wilderness. They are kin. They are not in the way of development – they are the way we developed.

We just forgot.

In forgetting, we lose not only ecosystems but stories, the myths that helped us understand our place within them. Among us Luo peoples, many folk tales speak of two brothers: the White Nile and the Blue Nile, who travel great distances before coming together in the city of Khartoum. These brothers have come to symbolize many things: land and water, nature and people, memory and forgetting. They intermingle, diverge and meet again, just as our lives cross and re-cross with the natural world. For those of us in the diaspora, no longer living in daily intimacy with these rivers or the stories that once shaped us, their meaning risks slipping away. But there is one truth to which the tale always returns: the two brothers cannot exist without each other.

My late father once told me the tale of Omondi, Otieno and the Hyena, when he was teaching me the message of interdependence.

Long ago, on the shimmering shores of Lake Victoria, lived two Luo brothers: Omondi, the elder, tall and strong as a papyrus reed, and Otieno, younger, quick-witted and sharp as a spear tip. Their parents had died when they were young, leaving them to rely on each other for survival.

The land was changing – rains had failed, and the earth had cracked under the sun's relentless gaze. Fish were scarce, and their small millet plot yielded little. One evening, as the orange sun dipped behind the acacia trees, Omondi said, 'Brother, we must go into the forest. Alone, we will not last another moon.'

With only a gourd of water and a woven basket, they set out at dawn. The forest was alive with the calls of birds and the distant laughter of hyenas. Deep inside, they found tracks – hoofprints and pawprints, but also the telltale signs of hyena: twisted, looping and clever.

That night, as they camped under a fig tree, the brothers were ambushed by a cunning hyena. The beast, sly and silver-eyed, waited until Omondi left to fetch water, then crept up on Otieno. Otieno, quick-thinking, climbed a tree, but the hyena circled below, snapping its jaws.

Omondi returned and tried to chase the hyena away with a stick, but the animal was fearless. It darted between them, separating the brothers. Omondi, strong but alone, could not outmanoeuvre the hyena's tricks. Otieno, clever but without his brother's strength, could not descend from the tree.

Hours passed. The brothers called to each other, voices echoing through the trees. They realized that only by combining their strengths could they survive. Omondi distracted the hyena, shouting and throwing stones. Otieno, seeing his chance, dropped down and used a thorny branch to prod the hyena from behind. Together, they drove the animal off. Exhausted but safe, the brothers embraced. They shared what little food they had and made their way home, vowing never to let pride or fear separate them again.

As the elders say, '*Jowi, odong gi jowi*' (a buffalo is never alone, it is always with others).

*

Folk tales about wetlands in the UK and elsewhere often begin with fear. That fear makes sense: without knowledge, wetlands can be dangerous.

Much of the British folklore that grew up around these places reflects the experience of people who moved into wetland areas and tried to drain or settle them – stories born from the difficulty and uncertainty of turning shifting, treacherous ground into cities. Jenny Greenteeth is the old water hag of British lore, a slimy, green-skinned crone with razor teeth and long, tangled hair. Parents used her name to keep children from dangerous waters, and the story lives on in the duckweed itself. As the Wisconsin writer and conservationist Tod Highsmith notes, 'passing by a marsh on a misty evening' can mean an encounter with 'Jenny Greenteeth, a sharp-toothed crone who pulls unwary wanderers into the depths and devours them'.[2] Under the lily pads lurks Jenny's silhouette, as she reaches out with long arms to drag her victims under.

Jenny is not alone in the wet world. Folklorists point out that she's a local face of a very old global archetype. The Yorkshire grindylow is practically her twin, a similar swamp monster snatching children into muddy ponds. In Finland the *näkki* hides in the water and tugs at toddlers' legs, and a Japanese *kappa* – a turtle-headed river sprite – will also seize and drown careless swimmers. And Australia has its bunyip, a forbidding swamp creature said to haunt the billabongs. As one folklorist observes, Jenny Greenteeth is 'a similar figure' to these other water-demons in world lore.[3] In every case, these hags and imps embody the wetland's most dangerous quality: unpredictable water. They personify our fear of drowning and the unseen depths beneath placid surfaces.

Communities used Jenny to personify danger, teaching respect for places where drowning or getting lost was a real threat. And although these legends are all about fear of what lies beneath the murky water, they are also charged with an odd reverence. Even as Jenny Greenteeth embodies dread, the bunyip of Aboriginal stories is called a 'wetland guardian', watching over rivers and billabongs. In Cairo, the frog-headed goddess Heqet presides over river floods, blessing the soil with fertility. By personifying the swamp, folklore reveals how our ancestors saw wetlands as places of power – capable of harm, certainly, but also vital to life.

*

When I speak of wetlands, I do not mean just one thing. They are not a single landscape or habitat, but a world of watery thresholds – marshes, swamps, bogs, mangroves, estuaries – each with its own personality. Each alive in ways we are only beginning to understand. What they share is a resistance to definition. Wetlands stretch, sink, sponge and shift. They blur the boundaries between land and water, between settlement and movement. They absorb the world rather than repel it. And yet, paradoxically, it is this very quality that has made them the foundation of some of the most enduring human civilizations.

One of my favourite places in Oxford is called Mesopotamia. It's a small ait – a river island – tucked between the upper and lower streams of the River Cherwell in University Parks. Taking that route to work adds at least 20 minutes to my walk, but that's part of the point – it invites a different pace. The slowing down allows something else to surface. The name, borrowed from the Greek *meso* meaning 'between' and *potamos* meaning 'river', is a nod to a much older geography, a wetland world that gave rise to some of the earliest cities humanity ever built.

The original Mesopotamia, now within modern-day Iraq, parts of Syria, Turkey and Iran, lay between the Tigris and Euphrates rivers. It's often called the 'Cradle of Civilization', with good reason. This was a land of marshes, floodplains and shifting waters. Contrary to the dry desert image often conjured in the Euro-American imagination, early Mesopotamia was a seasonally inundated wetland, pulsing with both peril and promise. Annual floods brought rich alluvial soil that turned otherwise arid land into a breadbasket, but the waters also needed tending, understanding and sometimes coaxing into patterns that could sustain life.

Here, the first cities took root not in spite of the wet, but because of it. The Sumerians, arguably the world's first urban people, became engineers because the wetland demanded it. They carved irrigation canals, designed levees and invented systems of measurement and record-keeping to monitor crop yields and water flow.[4] The cuneiform script itself was born from clay tablets, written with reeds taken from marshy banks – writing, quite literally, made possible by wetlands. Urbanism here was not a story of triumph over nature, but one of entanglement with it.

The city of Uruk, one of the earliest and most influential Mesopotamian city-states, arose in this marshy delta. With an estimated population of 40,000 as early as 3000 BCE,[5] Uruk was a marvel of infrastructure and ritual, featuring temples, granaries and public squares – almost all of it built from mudbrick, a technology made viable by the wetland's materials and climate. The very idea of a city – planned, walled, storied – emerged from the practical and spiritual necessity of living in watery terrain.

Recent archaeological discoveries in Eridu, one of southern Mesopotamia's oldest sites, have revealed vast and intricate irrigation networks dating back to the 6th century BCE.[6] These systems allowed for the controlled flow of water through seasonal wetlands, enabling stable agriculture even in years of climatic uncertainty.

The people of Mesopotamia didn't tame the land. They entered into a kind of dance with it. Their urban achievements – Uruk's ziggurats, Babylon's code of law, Sumerian cuneiform – were grounded in the landscape in which they lived. They used reeds from marshes to write, mud from riverbanks to build, and silt-laden waters to grow food. They built with what the wetland offered and organized society around the gifts and limits of water.

Their story is not unique.

Along the Nile, another rhythm shaped another remarkable civilization. Ancient Egypt flourished because of its relationship with water, but this relationship had a different tone. The Nile's floods were more predictable than those of the Tigris or Euphrates. Each year, the waters rose and then receded, leaving behind nutrient-rich silt. The Egyptians called their land *Kemet* – 'the black land' – a name inspired by the dark, fertile soils left by the floods. They embraced this annual pulse as a blessing, not a problem to solve.

The Egyptians developed early barrages and channel systems to expand the reach of these floods, transforming seasonal inundation into sustained agricultural productivity. Over time, the scale of their engineering grew. By the 19th and early 20th centuries, they were raising the Nile's water levels with complex hydraulic systems. And with the completion of the Aswan High Dam in 1970, Egypt's relationship with the river entered a new, modern phase, one marked by large-scale control rather than cyclical engagement. That dam brought with it profound changes: increased water storage, electricity generation, expanded agricultural lands. But it also came at a cost. The annual floods stopped and so did the natural deposition of silt. Soil salinity increased, while coastal erosion and waterlogging became persistent problems. In gaining control, something was lost, the delicate give-and-take that had defined the Nile's ecology for millennia.

Both Mesopotamia and Egypt illustrate a powerful truth: civilization doesn't arise in opposition to wetlands, but through relationship with them. When we talk about ancient societies, we often focus on palaces, pyramids and war. But what lies beneath those accomplishments is the subtle genius of water management. It is the story of irrigation ditches, not swords, that gives us a deeper insight into how we got to where we are now.

This pattern repeats across human history. Nearly every major city – London, Paris, Delhi, New York, Cairo – is founded beside a river, lake or wetland. Johannesburg, high on the South African plateau, is one of the few exceptions that proves the rule. Its rise came not from water but from gold. When deposits were discovered on the Witwatersrand in 1886, a mining camp exploded into a city almost overnight. From the beginning, Johannesburg was built on extraction: wealth pulled from deep underground, with water, food and labour brought in at great cost. That inheritance remains etched into the city's fabric – mine dumps, polluted groundwater and a landscape shaped as much by absence as by abundance. It stands in contrast to water-born cities, whose growth has long been fed by currents of trade, story and sustenance.

We're taught to associate civilization with dryness, with roads, bricks, clean lines and separation from the elements. But what if that's only one story? What if civilization can also mean being in relationship with seasonal floods? What if the true markers of advancement lie in our ability to listen to the land rather than trying to dominate it – to build a rhythm with it rather than work to discipline it as though it were somehow rebelling? Wetland peoples – like Luo peoples, the Marsh Arabs of Mesopotamia, or the fisherfolk of Ghar el Melh – have long embodied this quieter kind of genius. It's not lesser. It's just less loud.

In contrast, industrial approaches to water often feel adversarial. We try to control it – canalizing rivers, draining marshes, straightening every meander into a manageable line. But in doing so, we lose more than just biodiversity. We lose the buffers that protect us from floods. We lose natural cooling systems that regulate climate. We lose the deep memory stored in wet soil and layered silt. And, maybe most importantly, we lose the chance to be in relationship with a landscape that invites cooperation, not conquest.

The ways in which we speak about water and food, and our feelings about them, are not just cultural flourishes. Our language holds ecological memory. It encodes ways of living with change, of noticing nuance. In a sense, our stories are as important as our structures. What survives a flood is not just the house, but the habit of naming what matters.

Wetlands have always required humility. They demand attention.

They remind us that the ground may not always be firm, and that the way forward may not be straight. Yet, for that willingness to pay attention, they offer abundance in return. Abundance not only of material goods – clean water, fertile soil, fish – but of time, culture and imagination. They are not the leftover spaces on a planner's map. They are the reason cities exist at all. Wetlands are not margins – they are foundations.

This is not only a truth of the past. Cities today are rediscovering what happens when we work with water rather than against it. In the Polish city of Łódź, for instance, the Sokołówka River had been canalized during the industrial boom of the 19th century. Straightened and confined, it could no longer slow floodwaters, filter pollution or cool the surrounding neighbourhoods. A recent restoration project reversed that logic: the river was allowed to meander again, its banks softened, ponds and wetlands reintroduced. These new watery spaces absorb heavy rains, clean the water as it passes, and bring greenery and shade into a densely built city. It changed the way residents moved around the city and the air feels different.[7] It is a small reminder that in a time of climate crisis and urban sprawl, straight lines and hard edges do not serve us well. When we allow water to curve and settle, the land recovers some of its memory – and in that recovery, cities do too.

*

In Ghar el Melh, a small coastal town on Tunisia's northern shore, the wetlands breathe with history. Once known as Rusucmona – a significant Phoenician and later Roman port – this place has worn many names, and like the marshlands that cradle it, it has never stood still. By the 17th century, it had become a fortified haven for Andalusian pirates, its harbour filled with corsairs and cross-currents of culture. And now, in a quieter century, Ghar el Melh has metamorphosed again, from what was a site of conquest into one of care.

Its tidal lagoons and salt marshes are no longer simply geographic features. They are living systems that hold memory in water and sediment. In 2007 it was designated a Ramsar Wetland City, an honour that speaks not just to ecological abundance but to the commitment of the people who live there.[8] Fisherfolk, schoolchildren, historians and ecologists work side by side, not in nostalgic reconstruction, but in a shared effort to keep the wetlands alive and evolving.

Ghar el Melh shows us what it means when a city leans into its wetness, when it doesn't flatten or reclaim, but rather listens and adapts. Traditional fishing techniques still respond to the seasonal pulse of the tide. Migratory birds arrive, as they have for centuries, welcomed by marshland sanctuaries. Eco-tourism grows alongside heritage education, rather than in conflict with it. And in the surrounding agricultural fields, water use for irrigation has dropped by 44 per cent, even as crop productivity has risen by 66 per cent, evidence that wetland-based innovation can be as pragmatic as it is poetic.[9]

Here, the wetland is not treated as an obstacle to development but as its very foundation and a partner to be understood. In Ghar el Melh, the boundaries between land and sea, past and present, human and nature, are not erased but softened. And in this softness, something enduring takes root: a form of urban life that remembers how to coexist, not dominate.

Half a continent away, in Liverpool in the United Kingdom, a different kind of wetland story is being told. The River Mersey once flowed through Liverpool like a scar. It carried the waste of industry, the soot of empire and the heavy sediment of loss. For generations, it was a boundary – between counties, classes and communities; a river turned repository, its waters darkened with the burdens of the Industrial Revolution. But rivers, like wetlands, rarely accept finality. They move, they shift, they return.

Today, the Mersey is being listened to again. Along the Speke and Garston Coastal Reserve, wetland habitats are being reintroduced – tidal marshes, estuarine pools, bird corridors – not as restorations of a particular past, but as gestures towards a future where the river becomes an active participant in the city's renewal.

The Mersey is also a place of spiritual significance. British Hindu peoples perform rites along its banks, offering flowers and flames into waters that, for them, carry echoes of the Ganges.[10] It is also a corridor for birds on their own pilgrimages – swans, oystercatchers and migrating waders – following ancient routes across hemispheres. In this city, where docks once swallowed whole ecosystems, the river now makes space for return. It is an ecological and civic restoration, a shared effort, between local authorities and communities, to reclaim relationship – to water, to land, to one another.

As important as citywide policies and planning frameworks is the commitment of individuals, the people who show up quietly to do the day-to-day work of wetland care; those who restore, protect and advocate for wetlands not only as ecosystems, but as shared homes and sources of meaning. Their stories are an important reminder that while grand rewilding visions inspire us, it is often small, deeply attentive acts that change landscapes.

※

In Sussex, the wetlands aren't vast deltas or roaring rivers, they are tucked-away places, steeped in geology and time. Fran Southgate, Wilder Landscapes Advisor for the Sussex Wildlife Trust, describes one valley with reverence. She says, 'It's kind of like the land that time forgot.'[11] There are chalk springs bubbling through rare alkaline and acidic soils, creating fens so rich in life that it feels almost prehistoric: red lichens, dripping ferns, fallen trees tangled in vines. A place where the past hums in every stone and splash.

Fran's work mimics natural flood management. She doesn't try to force water into submission – she helps it slow down, settle, absorb. In these wetlands, she says, you can literally watch change unfold second by second. Reedbeds filter pollutants through microbial action in their roots; floodplains cradle excess rain and release it gently. The wetland teaches us patience and resilience through its dynamic ecosystems. Fran's reflections strike a chord. In a time where everything urges speed and certainty, wetlands offer a different kind of wisdom. One that says: accept change, anticipate flux, make space for softness.

In Sussex, we've lost at least 8 per cent of our original wetlands. Maybe more. And yet, what remains continues to teach. These are not wasted spaces, as some might claim. They are spaces of repair, of slow recalibration for both the land and us.

*

Thousands of miles away from Sussex, in the subtropical vastness of the Florida Everglades, Diana Umpierre is fighting a battle for recognition and renewal. As an organizing representative for the Sierra Club, she works alongside communities, particularly the Miccosukee and Seminole tribes, who have called these wetlands home for generations.

The Everglades were once dismissed as 'swamp', an obstacle to American progress. Beginning in the 1830s, they were dredged, drained and divided, until Lake Okeechobee became a kind of unnatural pressure point. When storms come, there's nowhere for water to go but outward – east and west – carrying with it toxic algal blooms, nutrient overload and threats to human health. Yet amid this engineered fragility, Indigenous people continue to steward what remains.

Diana speaks with quiet urgency about tree islands, slight elevations within the Everglades where ceremonies are held, food is cooked and the dead are buried. These are sacred places, but modern water mismanagement has flooded many of them, rendering them inaccessible or lost. 'How do we make sure we don't keep making the same mistakes?' she asks.[12]

For the Miccosukee and Seminole peoples, the Everglades are not a wetland to be protected in the abstract. They are home, ritual, history. To damage them is an ecological loss and also a form of cultural erasure. Wetlands store water and also memory. They are living archives, rich with the sediment of story, song and seasonal return.

*

Despite centuries of disruption – from colonial re-engineering to global market pressures – wetland communities continue to cultivate profound, lived relationships with water. These are more than vestiges of the past, they are adaptive, evolving ways of life that persist in the face of adversity.

The ecologist Mordecai Ogada and the mangrove scholar Dan Friess offer vivid reminders of this continuity. Ogada, reflecting on a proposed agro-industrial factory near Kenya's Nam Lolwe (Lake Victoria), told me about how multinational developers had promised jobs and piped water to the Luo communities living near a particular part of the lake. But what they didn't grasp was that water from a tap is not the same as water from the lake. Blocking access to the shoreline ruptured something more than geography – it fractured a spiritual relationship. 'They made the mistake of blocking people from the lake,' he said. 'And it all went pear-shaped.'[13]

The Luo people's response was not merely resistance to development – it was an assertion of a worldview. As Ogada put it, 'We are engineering solutions to problems that people who live with wetlands do not necessarily see as problems.' To them, the lake is not a resource to be extracted, but 'a jewel to be revered ... to be lived with'. In this worldview, wetlands are not inert landscapes awaiting transformation, but active kin to be stewarded, honoured and protected.

Dan Friess offered a different but resonant perspective from Southeast Asia. Quoting the Thai fisherman Mad-Ha Ranwasii, he shared the phrase: 'If there are no mangroves, the sea doesn't have any meaning. It's like a tree without roots.'[14] Mangroves, like freshwater wetlands, anchor culture as much as they anchor soil. Their resilience – starting life as fragile seedlings in volatile tides – mirrors that of the communities who rely on them.

These perspectives remind us that cultural continuity is more than just looking back with nostalgia to a time gone past. It is about the values that underpin the way in which the land is lived with and related to. Ways of seeing that are tuned to interdependence rather than extraction. The proposed factory near Nam Lolwe would have required materials to have been shipped in from afar and would have generated new ecological risks. But it was the spiritual rupture – the severing of people from their lake – that galvanized collective action. A holistic relationship with wetlands is not focused just on ecological concerns, but on sustaining life in all its forms: it's about ritual, livelihood, imagination. It brings to mind the way William Wordsworth, in his poem 'Lines Composed a Few Miles above Tintern Abbey, On Revisiting the Banks of the Wye during a Tour. July 13, 1798', described the sound of water years later, hearing it not only as a river but as a thread of continuity between the land and the self, something that remains alive inside you even when you are far away.

> *Five years have past; five summers, with the length*
> *Of five long winters! and again I hear*
> *These waters, rolling from their mountain-springs*
> *With a soft inland murmur.*[15]

Today, wetlands face mounting pressure: to be drained, farmed, built over. Where this has happened, communities have not only lost income but time. Time for festivals, time for storytelling, time to sit with the lake. The loss of rhythm is as devastating as the loss of land.

But in places where cultural continuities hold, wetlands still hum with relationship. And that may be our best hope. Not conservation alone, but conversation – between people and place, past and present, utility and reverence. To damage wetlands, then, is not just to threaten biodiversity. It is to erode a culture's ability to remember itself. It is to break the link between past and present, to interrupt the conversation between people and place. Wetlands are not marginal. They are mnemonic. They remind us who we are, and how we came to be.

What if we stopped seeing wetlands as problems to be solved, and started seeing them as teachers? Not only for what they do – filter water, absorb floods, store carbon – but for how they do it. They move slowly. They make room. They refuse straight lines. They remind us that flourishing often requires softness and sway.

The 'wobbly' Thames at Earth Trust showed me this. A river allowed to meander invites life. It cools the land. It stretches its limbs. It doesn't push forward – it settles in. This is what cities need. Not control, but conversation. Not rigidity, but responsiveness.

The best wetland projects today aren't only led by central governments, but also need to collaborate with communities. They are best shaped by stewards – not just scientists, but fisherfolk, artists, farmers and elders. People like Diana Umpierre in the Everglades, or Fran Southgate in Sussex, or the women in my ancestral villages who knew how to read the marsh grasses and follow the frogs. They didn't need data to know where the water would rise. They listened as they learned from the land.

In a warming world, we need systems that don't crack under pressure. The future won't be won by rigidity – it will be shaped by wetland thinking: absorbent, relational, able to bend without breaking. The wisdom we need already exists – in marshes, in mangroves, in the ways our ancestors lived with water, not on top of it. We didn't just adapt to wetlands. We shaped them too: gently, attentively, over generations. Papyrus was harvested, not razed. Canals were dug with the seasons in mind. The wetlands were never untouched, but they were not ruined. The relationship was not extractive, but evolutionary. We changed each other.

So yes, the future is uncertain. Sea levels are rising. Droughts are intensifying. But we are not without guides. Wetlands themselves are a blueprint. They teach us how to slow down, how to share, how to shape flexible systems. The wetland cities of the future will not turn away from water. They will turn towards it. They will curve and meander. They will hold complexity and celebrate it. They will be full of life. And life, as history reminds us, is rarely limited to humans alone.

Chapter Nine

The Hoofprint Beneath the City: Learning from the Food Systems of the Past

A strange, deep noise wakes me. It's low and loud, almost a roar, like something is hurting or lost outside in the dark. It comes again, louder, stranger.

I climb across my bed and press my nose against the cold window. There's nothing out there but trees and shadows. The noise comes again – drawn-out, aching. It sounds upset but also familiar.

Then I realize: it's Rosetta, our family cow. She's mooing, but not the soft, friendly sound she usually makes when she wants a scratch behind the ears. This moo hurts.

Even though I know I'm meant to be asleep, and I'll probably get into trouble, I pull on my gumboots and head quietly for the kitchen.

I'm surprised to see the corridor lights on. I freeze. Then I hear footsteps, and before I can hide, Mummy rushes in from outside. She doesn't scold me. Something must be wrong. She goes up to the attic and comes back with old clean towels. She flicks the kettle on without saying a word and disappears out the door again.

Something is very wrong.

Rosetta's moo changes tone. It's longer and deeper. I don't dare follow Mummy outside, but I can't go back to bed either. Every bit of me wants to be near Rosetta. To see if she is OK.

Rosetta is our family's cow but I'm the only one who knows her real name. I gave it to her when Oloo, our gardener, was teaching me how to milk her. He taught me milking songs too but it seemed strange to sing a song to someone without calling their name and so I named her Rosetta. Rosetta is the best person to go on a walk with. She is curious, like me. We explore. Oloo says I should keep a rope around her neck to help guide her but I don't need it. We decide together where we want to go and what we want to see.

Oloo agrees that Rosetta is very clever. When her stomach hurts, she knows just which plants will help her feel better. He tells me this in a way that makes it feel like a secret about how animals carry their own knowledge, a wisdom most grown-ups would never believe. Rosetta shows me that kind of hidden knowing, and being trusted with it makes me feel older.

Mummy rushes in again, switches on the kettle, and I can't hold it in any more.

'Is Rosetta dying?'

She pauses just long enough for my stomach to tighten. Her eyes are wide and distracted.

'Just stay here,' she says quickly, and hurries back outside without another word.

I stand there, frozen. My mind spins, wondering what's happening to Rosetta. Then a new thought hits me – what if Rosetta's hurt? What if something awful is happening, and nobody will tell me? The kettle's steaming, but I barely notice.

I'm just about to rush outside when Mummy finally reappears, calmer now. She sees my face.

'No, Rosetta's having a baby.'

I burst with relief at the news. An emotion of pure joy washes over me as my mind races ahead to consider my new playmate in time for my ninth birthday. I search my mother's face to share in my excitement but her face is scrunched up like a used tissue. Something's not right.

'What's wrong?' I say in confusion.

'It's a boy,' she says, like it's bad news.

'Huh?'

'We can't keep boys.'

Everything I have been told about cows gets muddled in my head. Boys can't make milk. Boys grow into bulls and the only thing bulls are good for is serving cows. It's a lot of money to keep a boy until he grows into a bull, and anyway, Rosetta is our only cow.

My eyes sting. I feel sick. In our household, nothing is wasted. Rosetta's baby is going to end up on our plates.

*

We tend to associate cows with the countryside, but that's a historically inaccurate division. Just as wetlands have shaped cities, cows have shaped urban systems too. They were not banished by modernity – they were embedded in it. In cities like Nairobi or New Delhi, cows wander the streets, perhaps less frequently than they once did but still asserting their grazing rights with a remarkable ability to stop traffic.

But the way we understand cows in an urban context reflects something much wider about our food systems in cities. Their disappearance from daily life is part of a broader pattern: as cities grew, food was pushed further and further away, hidden behind packaging, processing plants and long supply chains. Where milk once came from a neighbour's cow, bread from a nearby mill, and vegetables from market gardens at the city's edge, now it all arrives from somewhere else, nowhere in particular.

Cows make this shift visible. They show us what happens when living, breathing parts of our food system are replaced with invisible, industrial networks. When we lose sight of cows in the city, we also lose sight of soil, grass, rain and the long entanglement of relationships that keep us fed. Food becomes something abstract, transactional – more product than process. And so, the question of cows in cities is also a question about how urban life severs us from the sources of our sustenance.

To think about cows is to think about what it would mean for food to be close again: not just geographically close, but close in relationship. What would cities look like if our systems of eating once again acknowledged that food comes from somewhere – and from someone, or something – living? That intimacy with food – its sources, its rhythms, its responsibilities – is something many urban systems have forgotten. But in other parts of the world, it remains central to how life is organized.

As a Luo woman, it makes perfect sense to me that one would prioritize cows in one's life. Traditional Luo homesteads are built around a *kul*, a circular enclosure where the family's cattle are kept in the heart of the compound. If you flew a drone overhead, you'd see this *kraal* (an enclosure for livestock) encircled by a band of open space, and beyond it, a ring of round homes. The architecture of these homesteads centres on two pillars of Luo life: cows, housed at the core, and the matriarch, whose home always faces the main gate.

This spatial logic isn't limited to traditional homesteads. Even in more permanent and fortified Luo settlements, such as Thimlich Ohinga, a UNESCO World Heritage Site in Migori County in Kenya, the centrality of livestock is evident. These dry-stone enclosures, some estimated at over 500 years old, reflect not just defensive strategy, but a worldview: a settlement in which animals, particularly cattle, are essential to both security and social organization. It's easy to forget that even in cities, spatial design once centred on animals too. The placement of kraals in

Luo settlements mirrors, in a sense, the cow-strewn commons of early London. Different geographies, but shared logics – of proximity, care and cohabitation. What we've lost isn't just the animal itself, but a whole way of ordering life around shared need.

Over the past 10,000 years, cattle and humans have shaped each other's evolution in profound ways. As humans selectively bred cattle for traits like docility, endurance, milk production or meat yield, cows began to physically and behaviourally adapt to our rhythms. The adaptation ran both ways. Perhaps the most striking example is the development of lactase persistence in humans – the ability to digest milk into adulthood. While most mammals lose this capacity after weaning, certain human populations – such as the Maasai peoples of East Africa, Norse peoples of Scandinavia, and a subset of South Asian pastoralist groups, notably the Toda and Gujjar – developed genetic mutations that kept the lactase enzyme active for life. In effect, our bodies evolved to match our way of life. Milk both nourished and rewired us. These adaptations are not universal, which makes them even more remarkable: they are living evidence, in our very cells, of long, intimate entanglements with cows. Our shared evolutionary path is inscribed not only in ancient myths, but in our DNA. We didn't just domesticate cows. In a very real sense, they domesticated us too.[1] If our bodies have evolved alongside cows, then what happens when that relationship is interrupted? What other kinds of cultural memory – or even biological inheritance – begin to atrophy when we move too far from the animals who once shaped our survival? These are questions about animals and about food. When we lose the daily closeness to the sources of our milk, meat or butter, we also lose knowledge of season and soil, and the skill that make those foods possible. Food becomes a product, stripped of the relationships that once bound us to land and animal alike.

While it might be a stretch, at least based on current archaeological understanding, to describe Thimlich Ohinga as a 'city' in the conventional sense, it nevertheless shares something profound with other urban centres around the world: a past shaped, in part, by cows. One such place, surprisingly, is London, where the rhythms and needs of dairy cattle once sculpted the city's infrastructure, economy and daily life.

*

From the 18th to the early 20th century, cows in London were not an unfamiliar sight. In fact, they helped shape what we now think of as modern cities – places like London, Oxford and beyond. Their presence wasn't just quaint or curious; it was infrastructural. Just like Port Meadow in Oxford, where cattle still graze freely today, peri-urban commons across the UK remain living reminders of our shared ecological heritage with cattle. On London's leafy fringes – Wimbledon Common, Richmond Park and others – cows once roamed under communal grazing rights that supported poorer households and seasonal herders alike. Chiswick, today known for artisan bakeries and yoga studios, quite literally means 'cheese farm'. The name's Old English roots (*Cēsewīc*), remind us that the urban-rural divide is newer than we think. Even the most polished postcodes once smelled faintly of milk.

Urban commons and green spaces were lifelines, places where keeping a cow meant securing milk, resilience and a kind of closeness to nature that urban life rarely permits. As enclosure and development crept across these commons in the 18th and 19th centuries, access dwindled, and as it did so, the social contract between city and countryside began to fray.

That thread hasn't broken entirely. In Cambridge, a medieval tradition still survives. Every spring, bulls and heifers are herded onto Midsummer Common, Coe Fen and Sheep's Green – green lungs threaded through the heart of the city. The cows bolster biodiversity, cut grass management costs and draw locals and tourists alike with their quiet bucolic presence. Yet even this centuries-old practice now hangs by a thread. Local councils are considering cuts to the city's out-of-hours 'pinder' service, the team responsible for rescuing cows that fall into the river. Without that care, farmers warn, they'll be forced to withdraw their herds. Public outcry saved this practice in 2021, but without human voices to champion the case for cows in the city, a tradition that has survived for centuries could be undone by a single line item in a municipal budget.²

Similar stories were to be found in the United States, though with a distinctly celebratory tone. In cities like Boston, where urban dairying was once a vital part of daily life, cows shaped both infrastructure and eating habits. By the late 19th century, the city's demand for fresh milk had given rise to a network of small-scale urban dairies, tightly woven into neighbourhoods and supply routes. These dairies were more than back-end operations – they became part of the public experience of food. In Boston, 'dairy lunch' rooms, cafés supplied by their own farms, flourished. One, the Oak Grove Farm café, boasted an 800-acre farm and 150 cows delivering fresh milk, lettuce and tomatoes directly to urban tables. These spaces created an early 'farm-to-city' connection, a direct link between soil and sip, much like my own closeness to Rosetta.³

*

Behind our urban commons in England lies another thread, largely forgotten, yet integral to the city's fabric: the drovers. The drovers were a skilled, hardy class of rural workers – part herder, part navigator, part negotiator. Often travelling vast distances, they were responsible for keeping herds together, healthy, and moving steadily through unfamiliar terrain and growing urban congestion. Many were Welsh, some were Scottish and others English farmhands, and they knew the routes by heart. They wore broad-brimmed hats and carried long sticks or goads for control and for communication, shaping the movement of animals with gesture and rhythm. Drovers needed wit, stamina and deep animal knowledge. They dealt with tolls, landowners, weather, disease and thieves, all while managing the unpredictable will of hundreds of beasts. In a pre-industrial city, they were vital logistical experts, living maps in motion.

If cows were the muscle of the urban food chain, then drovers were its nervous system. From medieval times through to the 19th century, tens of thousands of cattle made the long journey on hoof into London each year, destined for Smithfield Market, the throbbing heart of the city's meat trade since at least Edward II's charter in 1327. These weren't anonymous shipments. The movement was physical, loud and visible: animals gathered in outer districts – Islington, Holloway, Knightsbridge, Paddington, Mile End – and were herded through the city's arteries, often under cover of night to avoid traffic. A spectacle of living infrastructure. You'd hear it before you saw it – the lowing of animals, the barked instructions of men, the clang of the milkmaids' pails, the rhythmic clatter of hoof on cobblestone. London became a city that was built around the arrival of animals.

This was a time before refrigeration, before railways, before abstracted supply chains. By the late 18th century, some 100,000 cattle and 750,000 sheep were being driven into the city each year. The scale is almost hard to fathom. These animals reshaped more than just menus, they carved paths into the city itself. Roads widened, inns and holding pens sprang up along key routes, markets expanded, and entire neighbourhoods adjusted their rhythms around these seasonal influxes. Drovers weren't simply herders. They were facilitators of a peri-urban ecosystem, sustaining connection between land and labour, field and fork, body and city; a kind of choreography between humans and animals that left hoofprints not just in mud, but in the very structure of the city.[4]

Cows shaped more than the rhythms of city life, they also shaped its architecture. As the Georgian era ushered in rapid urban expansion, the grazing commons and pastureland that once fringed the city gave way to neat rows of terraced housing. But the cows didn't vanish – they adjusted. With fields paved over and land costs climbing, cattle were folded into the crevices of the city itself. In working-class districts like Whitechapel, Bethnal Green and Camden, urban cow-keepers kept small herds in their backyards, often near breweries and distilleries. This was a strategic placement.[5]

Those industries produced exactly what cows needed: feed. Spent grain and mash, the dense, nutrient-rich leftovers from making beer and spirits, were sold cheaply and in bulk to local dairies. Cow-keepers made opportunities from waste. For small-scale cow-keepers who couldn't afford to transport feed across long distances, proximity to these by-product goldmines was essential. It created a kind of urban symbiosis: the brewer brewed, the cow-keeper fed and the milkman delivered. As traditional grazing land was eaten up by bricks and rent prices, these informal ecosystems kept cows in the city just a little longer, and as part of a fully functioning urban food system, stitched into the daily life of the city.

Of course, keeping cows in the city came with consequences. In the 18th and 19th centuries, London streets were paved not with gold but with something far more pungent. Cow dung joined horse manure, rainwater run-off and refuse to form a kind of viscous sludge that coated the streets and seeped into every crevice. It was such a problem that crossing-sweepers, often children, were hired to clear paths through the muck for pedestrians. In response, Georgian builders quite literally raised the stakes: homes were designed with elevated entrances and proud front steps. These weren't just the aesthetic embellishments we tend to view them as today, but protective buffers. Raised thresholds helped stop animal waste, floodwater and filth from washing into parlours and cellars. Many entrances were flanked by boot scrapers, so visitors could remove manure and mud before stepping inside. What now looks like elegant Georgian charm was, at least in part, a response to living in close quarters with cows.[6]

Nowhere is this architectural adaptation clearer than in the mews. Originally built as service lanes behind grand townhouses, mews were designed for stables and storage, but in many neighbourhoods, they quickly became home to small-scale dairies and cowsheds. The proximity to households meant that milk could be delivered fresh each morning, straight from cow to doorstep. These were working spaces, often cramped and cobbled, filled with the warm scent of hay and the soft shuffle of hooves. Today, of course, London's mews are the preserve of millionaires and design magazines. It's hard to imagine now, as one peers through sash windows at polished interiors, that these cobbled lanes once echoed with the lowing of cows and the clang of milk pails. If only the estate agents knew. But forgetting is easy in a city that rebrands itself by the decade. The animals disappear. The dairymen retire. And we inherit the architecture with no memory of the labour, the lives or the relationships that made it necessary.[7]

Behind the elegant facades of Georgian and Victorian London, milk was being produced in real time. Mews courtyards held small herds, typically four to ten cows, tended by Welsh and rural English cow-keepers who brought with them generations of herding knowledge. Many of these dairymen were descendants of drovers, families whose ancestors had walked cattle across the country to market and now found themselves in a new kind of pastoralism that marked a contrast to the open fields it once took place in.

Welsh farmers, in particular, played a significant role. Known for their hardy hill cattle and strong dairying culture, they helped keep the city's milk supply alive – sometimes literally. Cattle would be brought in from rural Wales or the English Midlands by foot, canal and, later, by rail, settling in mews dairies to serve the surrounding neighbourhoods. These were micro-economies nested inside residential streets: cowsheds, hay lofts, milking stalls and the clink of early morning deliveries echoing off brick walls. A city block wasn't complete without its dairy. Often there were two. There were milk deliveries up to four times a day. It's easy to forget now, but milk in this era had to be fresh because it couldn't yet be refrigerated. Urban dairies weren't a quirky lifestyle choice, they were vital infrastructure. Milk was walked across yards, not trucked in from miles away. Children fetched it in pails. Women skimmed cream from glass bottles. Hygiene could be dubious, yes, but proximity was everything. And for families who couldn't afford the luxury of regular meat, milk was nutrition, economy and survival.

In some mews, the dairies persisted well into the 20th century, slowly replaced as pasteurization laws, refrigeration and the supermarket supply chain rewired the city. But remnants linger. You can still find odd barn-like doors and worn iron rings set into wall. Troughs that were once there for cows to drink water have been repurposed as planters. These are all architectural fossils of the cows that once called these streets home.

Today, mews homes command staggering prices, marketed for their charm and seclusion. But beneath the Farrow & Ball paintwork and reclaimed wood flooring lies a quieter truth: these were once the soft centres of the city's milk supply. Livestock lived here. Children knew their names. And for generations of Welsh, English and Jewish cattle-keepers, the mews was both workplace and lifeline.

If only the cows could see what became of their stalls.

One such story, perhaps the clearest in architectural and social terms, belongs to Welford Dairy. Its stalls, yards and even carved cow heads still linger in West London's red-brick facades, bearing silent witness to a time when milk production was hyperlocal, and urban life was interwoven with the rhythms of cows. Paddington may seem an unlikely place to begin a tale of pastoral infrastructure, but in the mid-19th century, it was still fringed with fields and cowsheds. That's where the Welfords began.

If you walk down the quiet stretch where Elgin Avenue meets Shirland Road in West London, you might pass a handsome red-brick building without giving it a second thought. But look closely, and you'll see cow's heads carved into the masonry, little architectural ghosts of a city once shaped by the milk it drank. This was once the Welford Dairy: a landmark of Victorian ingenuity, and one of the clearest examples of how London's development was, in part, designed around and shaped by cows.

Welford Dairy began modestly in 1845. Richard Welford set up at Warwick Farm near Paddington, then a semi-rural edge of the city. His family kept cows there, grazed them on open fields, and sold fresh milk to local households – a classic urban cow-keeping model. But London was already on the move. The Great Western Railway arrived. The fields began to shrink. And Paddington, once pasture, began its march towards bricks and mortar.

Rather than resist this change, the Welfords adapted. They began to move their cattle further out – to Oakington Manor Farm in Wembley and other green-edge sites – but they held on to the city by building a vast, purpose-built dairy complex in the heart of what was becoming Maida Vale. Completed in the 1880s, this new dairy was more than a milk processor: it embodied a changing urban relationship to food. Over one hundred horses delivered milk across the city, connecting rural supply to metropolitan demand with astonishing speed. The operation employed hundreds of workers and included workmen's flats on site, an early example of industrial and residential integration. Milk wasn't simply something people drank. It shaped where they lived and how they worked.[8]

Ghost hoofbeats echo
Old walls remember the cows
milk has slipped away

Welford's rise also tells the story of how land use in London shifted. As fields disappeared and housing density rose, small backyard cow-keeping operations (common in places like Whitechapel and Bethnal Green) gave way to more centralized, industrial-scale processing. The dairy at Elgin and Shirland became a node in a new kind of network, one in which cows were increasingly kept out of sight, even as their milk was brought ever more efficiently to the door.

The business was successful enough to catch royal attention. Welford's milk was supplied to Queen Victoria and the Prince of Wales. This was milk that symbolized prestige, hygiene and modernity. And it was this very image – of clean, white, mechanized milk – that eventually pushed cows even further from the city. As pasteurization laws tightened and refrigeration advanced, the rationale for keeping even outlying herds near London began to weaken. By the early 20th century, the cow was disappearing from the daily life of the city. But for a time, thanks to dairies like Welford's, she was very much at its centre.

Today, the old Welford Dairy building is still there. The cows, of course, are long gone. The cobbled mews that once echoed with hoofbeats now house cars, flats and boutique workspaces. It's the kind of street that estate agents describe as 'tucked away' and 'full of character', rarely noting that its character was forged in the hard, rhythmic labour of urban cow-keeping. If the carved cow heads remain, it is as a kind of architectural wink – proof that once upon a time, cows weren't just tolerated in the city. They were depended on.

Welford's story reminds us that cities were once designed around animal life. Milk was once a local product. Food systems were intimate, embodied and very much in motion. The disconnection we feel now – from where our food comes from, from the animals who help feed us – was not inevitable. It was engineered. Brick by brick. And perhaps it can be dismantled too.

*

Across the Atlantic, another imprint of urban cow-keeping survives – this time in the bold flourishes of early 20th-century American modernity. In Brooklyn, New York, Sheffield Farms – once one of the world's largest milk companies – built a monumental bottling plant. The former Sheffield Farms milk bottling plant still stands prominently at 1368 Fulton Street, between New York and Brooklyn avenues in Bedford-Stuyvesant, now called Restoration Plaza, its terracotta cow heads and milk-bottle reliefs serving as enduring reminders of its bovine heritage.[9]

Sheffield Farms didn't stop at infrastructure; they mastered spectacle. At the 1939 New York World's Fair, their marketing reached a new peak: Borden's 'Elsie the Spokescow' – a real Jersey heifer charmingly named 'You'll Do, Lobelia' – performed daily atop a rotating milking platform called the Rotolactor. Millions watched in awe as machine and mammal merged into a vision of hygienic abundance. It was part education, part entertainment – and unmistakably, a performance of intimacy between urban dwellers and the cow behind their glass of milk.[10]

*

Welford's rise and eventual dispersal mirrors a larger shift: from cows as part of the urban rhythm to cows as distant abstractions. As cities expanded, cows were pushed further out. Out of the commons. Out of the mews. Out of sight. Today, in most European cities, we rarely meet the animals whose milk we drink or the systems that sustain that relationship. And with that distance comes misunderstanding: of the animals themselves, and of the food systems they supported. Once, cows stood at the heart of a city's social and ecological infrastructure, but now they are often treated as a problem to be solved. And not just in urban areas. Even in the countryside, where you might expect deeper ties, they've become symbols of conflict – caricatured as methane machines, blamed for climate change, land degradation and inefficiency. But is it really the cow we should be questioning, or the systems we've built around them?

The rise of high-tech industrial farming has shifted the relationship dramatically. Landscapes are cleared for monocultures. Chemicals are spread not in response to illness, but in anticipation of it. Animals are bred for speed and yield, rather than resilience or relationship. As cities severed themselves from their food sources, they outsourced meat and milk to an industrial system that prioritizes volume over connection. In this context, of course cows appear to be at the centre of a crisis. They've been abstracted from the fields, commons and city lanes where we once met them, and folded into industrial cycles that prioritize production over patience. We worry about emissions, and rightly so, but rarely ask how we got here. How can something we co-evolved with over thousands of years, something written into our genes, our stories and our languages, now be cast as fundamentally incompatible with our future?

There is, however, another way to see them. In regenerative agriculture, cows are no longer cast as culprits in the climate crisis or reduced to mere milk-and-meat machines. They're understood instead as ecosystem engineers. Their grazing patterns aerate soil, disperse seeds, fertilize land and help manage grassland health. When allowed to roam and graze in rotation, cows can actually increase biodiversity and carbon sequestration, restoring landscapes instead of degrading them. Like beavers in rivers or elephants in savannahs, cattle – when stewarded thoughtfully – become co-designers of resilient ecosystems.

Perhaps the question isn't whether cows belong in the future, but whether our future cities and food systems will make space for the kind of relationships that honour their role. What happens when we sever nature – in this case, food-producing animals – from our idea of civilization? We lose more than just a species or a farming method. We lose a part of ourselves: the softer, slower, more reciprocal ways of being that once sat at the centre of urban life and made feeding ourselves a collective, visible act.[11]

A more balanced future might not be one in which cities push nature ever further to the margins, but one in which we draw it closer again. Not as ornament, but as kin. Not simply as carbon offset, but as collaborator. Food is central to this. Too often the solutions offered for feeding cities in the future – vertical farms sealed in glass towers, underground farms lit with LEDs – treat food production as something to be hidden, something sterile and detached from the soil and the fabric of the city itself. These models may shorten supply chains, but they risk cutting us off even more from the living systems that nourish us.

I often think of Rosetta, as part of a lineage – a soft infrastructure of animal life that once tethered humans to the land, to one another, and to time itself. Among Banyankole peoples of Uganda, daily life is measured around cattle – milking time is not just a chore, but a marker of social rhythm. Time, in this context, is not an abstract pressure that drags people from sleep in the dark hours, governed by clocks divorced from seasonality. It is shared with more-than-humans. And perhaps this is one of the things cities most need to recover: a different sense of time. Modern urban life tends to be short-sighted, rushing towards growth, speed and novelty while forgetting that flourishing often comes from slowness. Animals can teach us this. So can trees. If we are willing to pay attention, a forest on the city's edge reminds us to take the long view: to design food systems and urban spaces that endure for generations, not just quarterly reports.

Perhaps what our cities need is not more steel and digital precision – but more cows.

Chapter Ten

The Long Game: Seeing the Wood and the Trees

A motley crew of us gathers around a tree guide at the Nairobi Arboretum – a crowd of those who have time on a mid-morning Tuesday. There are a few retired Kenyans, a cluster of expat wives whose husbands work for aid agencies, and me, with my daughter asleep against my chest. We are, for the moment, a small picture of leisure.

Our guide keeps us entertained, pointing out trees as if they were characters at a party. Who belongs here, who arrived uninvited, who carries a story from far away. When he pauses at a pine and calls it exotic, a Spanish woman frowns.

'Pines are not exotic,' she insists.

'Here they are,' someone replies. 'We're in Kenya.'

She still shakes her head. 'But pines are common.'

'In Spain,' another woman says, 'not here.'

The guide smiles patiently. 'Exotic means from somewhere else. You moved here nine months ago, right? Would you say you're Kenyan?'

She laughs, a little embarrassed. 'No.'

'For us, you are exotic,' he says. 'And so are pines.'

As we walk to the next trees, I hang back, still turning this exchange over in my mind. The guide was right: the woman had been looking at Nairobi as though she were still standing in Spain. To her, the pine was ordinary, familiar – she could only see it through her own lens. In that way, the complexity of the arboretum was flattened; she missed the patchwork of species gathered here, each with its own history of arrival. The guide's words unsettled her because, for a moment, she was the one made strange – othered in a way she was not used to.

And I wonder if that is how we often look at both forests and cities, through the narrow lens of what feels familiar, blind to the layered lives and crossings that make them what they are. These trees seem to say otherwise. Like cities, forests are never simple; they are made of difference, of species and stories that have always been here and others that have travelled far to take root.

As scholar Elizabeth Oriel points out, trees stand as proof that the boundaries we draw between nature and culture are thinner than we like to think.[1] They carry the history of human hands: the ambition that planted them, the neglect that almost lost them, and the care that has kept them alive. Some are here because of colonial experiments in speed and control; others because someone simply wanted shade. And yet, whatever the reason, they go on making air, softening light, cooling the city – supporting us whether we have been generous or careless. When we dissect the story of the forest and who was here first, these trees teach us a lot about belonging, and about how deeply our lives are braided together.

I drift a little away from the group, my daughter's head heavy against my chest, and find myself thinking about how the Nairobi Arboretum began. It's not ancient. Its development reminds us that forests are not always the wild untouched places we assume them to be. They can be deliberate and tended. It was planted in 1907 by a British forest officer,

Mr Batiscombe, as a kind of experiment. At the time, the Kenya–Uganda railway had just carved its way through what was then swampy savannah plains. The railway needed wood, fast-growing wood, to keep the steam engines fed and to build the city that was springing up around the tracks. Indigenous forests were seen as too slow to grow, so seeds were gathered from as far away as Australia, India, Mexico and New Zealand, and planted here on a dry, stony slope to see what might take root.[2]

That railway is the reason Nairobi became a city. It brought with it all the elements of a new urban centre – depots, markets, offices – but it also consumed trees as quickly as it built. The arboretum was meant to solve that problem: a test bed for trees that could keep pace with the ambitions of a new colonial city.

Over the decades, the place changed. Indigenous and exotic trees grew side by side. Later, when the city's attention turned elsewhere and the arboretum began to decline, it was revived by ordinary people who refused to let it disappear – a different kind of long game.

Looking around at these tall trunks and shaded paths, it's easy to forget that just over a century ago this was nothing but stony ground and scattered thorn trees.

*

The strange thing about travel is that no matter how much one may enjoy the new and unfamiliar, there is always a natural pull towards the familiar. That thing that makes us feel 'at home' even when we are far away. Global brands have learned to utilize this push and pull. Chains like McDonald's localize just enough to 'belong' in the countries where they operate, but keep things familiar so that travellers craving something known can step inside, knowing exactly what to expect, and finding comfort in that sameness when everything around them feels unsettling or just different.

For me, navigating these feelings on my travels has often come in the shape of plants and, more specifically, trees. One of my favourite urban trees is the jacaranda (*Jacaranda mimosifolia*). In the Nairobi Arboretum, and across the city in October just before vuli – the short rains – the city is cloaked in a breathtaking purple-blue. The trees' large trumpet-shaped blossoms first cluster and then carpet the streets, to the delight of residents and visitors alike, some of whom say it makes them feel as if the city is treating them like royalty. As a child, I was equally enamoured with the jacaranda seed pod: as it matures, the pod forms a hard woody shell, like a little clam (or, as I liked to think, a pair of castanets) holding the seed inside.

It is such a distinctive tree that it became a symbol of home for me. So when, on our travels, I began to see the same tree in other cities – Johannesburg, Mexico City, Lisbon, Los Angeles, New Delhi – I became intensely curious. What kind of chain was this tree, thriving in such different places? And the effect of these blooms is never only visual. Elizabeth Oriel calls this 'vegetal affect': the way plants generate emotion and even a sense of community. Nairobi's jacarandas, like those in Mexico City or Pretoria, are not just decorative. They create a season, an atmosphere: a rhythm of anticipation and delight that, for a few weeks each year, links strangers who look up at the same purple canopy.

Its origins lie in the same part of the world that gave us many now-global plants, including the potato and the tomato. The jacaranda is native to South America. Seeing jacarandas in so many places made me realize that the migration of trees to cities tells us a lot about how we develop these urban centres and how we make, or don't make, place for nature in them. Trees cross oceans, travel with people, and take root in places far from where they began. Jacarandas don't erase the local landscape, they add to it, blending histories. A jacaranda in Nairobi isn't the same as one in Mexico City, but both remind us that trees pay little attention to the boundaries we draw between here and there, or between what we think of as 'nature' and what we call 'culture'. They are evidence of how entangled those worlds really are: shaped by us, shaping us in return.

The jacaranda's journey is not natural migration but human-led transplantation: a botanical thread running through colonial networks, urban planning and cultural memory. These trees bridge nature and human culture, revealing both entanglement and empire. Jacarandas first reached Africa via Cape Town around 1830, introduced through the efforts of early horticulturalists such as Baron Carl Ferdinand Heinrich von Ludwig, a German émigré who settled at the Cape in the early 19th century and established one of its earliest botanical gardens.[3] Alongside institutions like Kew Gardens in England, these networks of plant collectors and gardeners set the stage for the jacaranda's global journey. From there, the tree began its odyssey, planted in Australian botanic gardens and colonial cities before being propagated into the outskirts of new settlements.

In Kenya, jacarandas arrived with Boer families who relocated after the Anglo-Boer War (1899-1902). When the British defeated the Afrikaner republics of the Transvaal and the Orange Free State, many Boer families were displaced from their farms in South Africa. To encourage loyalty and stabilize the newly annexed territories, the British administration offered some of these families land in what was then called the White Highlands of Kenya. They travelled north with whatever would help them recreate a sense of home, and among those things were jacaranda seeds and saplings. There is an irony in that choice. Just as travellers reach for the familiar in an unfamiliar place, these settlers sought to recreate a landscape they recognized, even as they were reshaping someone else's. The jacaranda became a way of softening new surroundings, a splash of South Africa transplanted to East Africa, and over time the trees came to symbolize not just prestige but also the enduring push and pull between the known and the strange that comes with being uprooted. Eventually, what began as a private gesture of comfort – planting a tree to make a strange land feel more like home – spilled into the public realm. Colonial administrators soon adopted the jacaranda in Nairobi, Nakuru, Eldoret and other high-altitude Kenyan towns, favouring its annual blossoms and ease of seed propagation. By the 1920s, jacarandas were being planted and managed as street trees and public ornamentals by colonial governments, solidifying their identity as urban markers of modernity.

Long before Nairobi, Pretoria in South Africa had already been made a home for the jacaranda. The first two seedlings were planted in 1888 in the garden of Jacob Daniël Celliers in Sunnyside, Pretoria, after he bought them from a travelling nurseryman named Tempelman. These 'parent trees' became the source of seeds for many others across the Transvaal. A few years later, a local nurseryman, James D Clark – better known as 'Jacaranda Jim' – saw their potential and in 1906 donated 200 young trees to mark Pretoria's 51st anniversary, lining the streets with what would become the city's emblem. But it was Frank Walter Jameson, another 'Jacaranda Jim', who scaled this up. Appointed Town Engineer of Pretoria in 1909, Jameson established a city nursery, and in 1911 embarked on a massive planting scheme: 40 miles of jacaranda-lined streets. By the time he left for Kimberley in 1920, over 6,000 trees were thriving, and Pretoria was well on its way to becoming the 'Jacaranda City'. Jameson carried this zeal with him, later planting thousands more trees in Kimberley – and even Nairobi, where his influence worked to define the colonial cityscape. These imported trees, planted to impose a familiar vision of order and beauty on an unfamiliar landscape, have left behind something far more complex: living markers of entanglement. Intended as instruments of control, they have instead taken root as reminders, of how places and people are shaped as much by violence and displacement as they are by care.

The history of how jacarandas came to Nairobi carries this tangle for me. Trees have a double life in cities. Planted as tools of control, they almost always grow beyond that intention, as they root themselves into the life of a place, creating something that can't be planned – a living network of shade, scent and memory. They gather people around them in ways that outlast the purpose for which they were planted. I love their October bloom, but I cannot forget that they arrived here through a colonial history that was deeply disruptive for families like mine. The two countries I now move between, Kenya and the UK, are tied together by that unequal past. And yet these trees also show me that, even within those legacies, meaning can take root. Over time, they have become as much a part of the city's memory as its skyline. As Oriel writes, urban trees are multispecies actors in placemaking: they shape aesthetics, memory and identity, even when alien to their setting.[4] Jacaranda blooms mark time and place – spring rituals, festivals, student superstitions in Pretoria, and the October delight of Nairobi – reminding me that resilience and beauty can grow out of even the most difficult histories. They have taught me to see cities themselves as ecological: not separate from nature, but made by it, just as much as by us. Ornamental and wild; belonging and foreign: wherever they bloom, these plants narrate entangled histories – colonial, botanical, social.

Across British colonial cities, jacarandas followed patterns of infrastructure and settlement. From Kew Gardens to botanic collections in South Africa, propagated seeds were distributed to colony gardens and transport networks. Their proliferation signals both aesthetic choices and imperial logic, that of cities shaped by trees sourced elsewhere.

Their survival, however, depends on care. Without community associations – like Friends of Nairobi Arboretum – or deliberate heritage protections, even iconic urban trees are vulnerable. Pretoria's public outcry over the felling of jacarandas speaks to how deeply they have become part of civic identity, and Nairobi's volunteers reviving a forgotten arboretum show how trees and people sustain one another. They remind us, too, that cities can be shaped by acts of displacement and still grow into places where different roots entangle. Like the people who now call these cities home, trees from elsewhere adapt, take hold, and contribute to the life of a place. Their long-lasting legacy shows us that for cities to thrive, they must remain open: made stronger by the presence of both the native and the newly arrived. This is why, in so many places, the jacaranda has become more than a tree. It is a witness. A reminder that belonging is never fixed, that it can be grown, season by season, into something shared.

Urban jacarandas thus stand as living proof that cultural belonging can be biocultural. Their annual blooms remind Kenyans and visitors alike that seasons endure, histories overlap and hybridity flourishes. The jacaranda is the same tree across continents, but each urban context localizes its meaning: in Nairobi, it marks vuli's arrival; in Pretoria, exam season; in Los Angeles or Lisbon, aesthetic cosmopolitanism. These layered meanings reflect both our best impulses and colonial legacies – care and erasure. Trees like the jacaranda silently carry those contradictions while offering shade, scent and spectacular colour – almost forgivingly.

The jacaranda's presence in Nairobi carries this lesson too: once planted, their work is slow and invisible. For years they look like nothing at all – just spindly saplings – but season after season they deepen their roots, and then one October they take over the skyline. They remind me that what flourishes in cities isn't born overnight.

*

There are two reasons, I think, why people are scared of taking on a project with the land. The first has to do with time. Real time. The kind that stretches beyond budgets, beyond political cycles, beyond the neat blocks of a school calendar. The kind of time that demands faith. It can take decades before anything looks like the vision you carry in your head.

The second reason is doubt. We're not always sure it's possible – that one person, or one family, could really shift the story of a piece of land.

We're taught to believe that transformation is the work of governments, of capital, of something bigger than ourselves.

But growing up, I saw with my own eyes that this isn't true. Forests can begin with dreams. My parents began with a bare, sun-baked piece of land in Kibos, once stripped by sugarcane, and, tree by tree, season by season, they coaxed it into a forest. Even when our lives were full of upheaval – changing houses, schools, jobs – the trees just kept going. They paid no attention to our timelines. They grew slowly, invisibly at first, until one day there was shade where there had been none, birdsong where there had been silence.

Watching that happen changed me. It taught me that forests, and perhaps all the things that really matter, are a long game. They are a daily act of patience, of trust, of putting something into the ground that will not fully belong to you, but to the future.

But perhaps the arboretum's greatest lesson is what happens when the relationship falters. After decades of growth, it began to decline in the 1970s and 1980s – not because the trees died, but because no one tended to them. No one maintained the bond between people and place. The site became insecure, underfunded, forgotten.

Then, in 1993, a volunteer group – Friends of Nairobi Arboretum (FONA) – stepped in with the Kenya Forest Service to resurrect it. Their efforts transformed the arboretum into the vibrant green refuge it is today. In the city, trees don't flourish from inertia. They flourish from care. The arboretum teaches us that trees need people as much as people need trees, especially in cities, where ecosystems depend on intentional stewardship.

And perhaps the most surprising example of this interdependence is Venice, a city we think of as built on water, but whose very survival depends on wood. Few marvels in human-engineered cities rival that of Venice. The city literally floats on an inverted forest: millions of wooden poles – oak, larch, pine, spruce, alder – driven deep into the lagoon's mud form the foundations of palazzi, bridges and piazzas. That wooden underlayer has supported Venice for over a thousand years.

Most people know Venice by its canals and frescoes, but the architecture owes as much to the forests of Northern Italy, Slovenia and Croatia as it does to stone. The city is a monument not only to artistry, but to the long view: wooden roots anchored in mud, bearing the weight of civilization above. A forest born of caregiving earth and generations of maintenance.

What binds these stories – of Nairobi's arboretum and Venice's submerged forest – is time and tending. Trees planted without care may sprout, but they won't endure. In both cases, success depended not on grand ambition alone, but on ongoing attention, patience and a willingness to invest beyond immediate returns.

It may seem strange to call the wooden piles of Venice a forest: after all, those trees were cut down long ago. But part of understanding our relationship with forests is accepting that engagement with trees doesn't always mean leaving them untouched. A forest can live on in a city, even when its form has changed, if that relationship is handled with respect.

My friend Rebecca, who is a forester, once told me something that surprised me: *if we don't use forests, we lose them*. At the time, it felt counterintuitive. I had grown up believing that the best way to protect a forest was to leave it alone. But as she explained, when forests hold no value in the lives of people, they become vulnerable. They can be cleared for sugarcane or for roads or for housing, because they are seen as empty space.

Using forests well – selectively, carefully and in partnership with the people who depend on them – is what gives them a future. In the Congo Basin, for example, parts of the forest that were managed under strict Forest Stewardship Council guidelines, with close involvement of local communities, made for a very different kind of forest management. Unlike surrounding parts of the forest where those communities were locked out, the areas that worked with local people were healthier and more sustainable. Some of the timber harvested from that forest travelled all the way to London, where it was used in the construction of Arsenal's Emirates Stadium. A football ground in a modern city may seem a world away from the tropical rainforest where those trees once stood, yet the connection is there: a forest woven, quite literally, into the fabric of urban life. Responsible use can allow forests to endure. When the relationship is one of care rather than extraction, the trees that are taken become part of a much larger, ongoing story – one where the city and the forest are not adversaries, but collaborators.

*

Growing up seeing a forest co-created from the ground up changed something in me. It softened the rigid categories I adhered to about what was 'natural' and what wasn't. When I witnessed how deliberate human care could regenerate land, I became more open to the idea that forests weren't only to be protected from human use, they could be protected *through* it.

Forests had long been part of my family's life in ways I hadn't fully appreciated. For example, we didn't use plastic toothbrushes when I was a child. Instead, we used small twigs cut from a plant widely known as the toothbrush bush (*Salvadora persica*). It's a shrubby tree found throughout East Africa, with branches that, when chewed at one end, fray into natural bristles. Not only is it sustainable, but the plant itself is antibacterial, helping reduce plaque, gingivitis and thus gum disease. Known as *mswaki* in Swahili, it has been used across cultures for thousands of years and is even endorsed by the World Health Organization for its oral health benefits. And it grew, quite unremarkably, around us – just one of the many unsung gifts from the forest that we took for granted.

When we had upset stomachs, particularly diarrhoea, my mother would give us charcoal. She didn't explain why at the time – only that it worked. Later, I would learn that activated charcoal binds to toxins in the gut and helps remove them from the body. The charcoal was often made from tree bark or wood offcuts, burned in a controlled way. It was one of the most ancient and effective remedies we knew, and it too came from the forest. The more I thought about it, the more I realized I hadn't been seeing the wood for the trees. Forests weren't just scenery or sanctuaries. They were pharmacies. Pantries. Toolkits. Places of healing and sustenance. Our lives were quietly entangled with them in ways we rarely paused to recognize. We used bark and roots to treat colds, leaves to make infusions for fevers, branches for fencing, and fruit for medicine. These uses weren't a threat to the forest – they were part of the relationship that sustained it.

*

I first met my foresting friend Rebecca when we were both students. We worked together on potato farms in Scotland as our summer jobs. It was an odd rhythm, inspecting root vegetables in rural Aberdeenshire, then returning to lectures and libraries in a post-industrial English city. But it was Sheffield in the north of England that cemented for me the fact that forests are a necessary part of cities.

When I first arrived, Sheffield's greenness surprised me. It wasn't something you expected from a city that once ran on steel and coal. But here were streets canopied with mature plane trees, old parks interlaced with walking trails, and entire neighbourhoods where the trees seemed to outnumber the buildings. I later learned that Sheffield has one of the highest tree-to-person ratios in Europe – an urban forest built over generations and fiercely defended by its residents.

During my time there in the 1990s, Sheffield was grappling with the remnants of its industrial past. Some parts of the city still bore the soot stains of former factories, but in between the grey were green eruptions. Old railway lines had been converted into wild corridors for foxes and cyclists alike. Residents campaigned not just to plant trees, but to protect the old ones, refusing to accept that urban infrastructure had to come at the cost of canopy.

This was the first time I really considered that trees in cities weren't just aesthetic flourishes or decorative extras. They were infrastructure – climate, culture and community rolled into one. And I began to see links between this and what my parents had created in Kibos. In both places, trees cooled the land, drew in birds, softened the light. In Kibos, the temperature under the forest canopy could be a number of degrees cooler than in the sun-scorched sugarcane fields nearby. In Sheffield, summer afternoons under the chestnuts and limes felt like a kind of refuge.

Across the world, cities are waking up to the power of trees. As we saw in chapter seven, nearly a third of Singapore, one of the most densely populated countries on earth, is covered in urban greenery. It's not by accident. In response to rapid urbanization, Singapore's planners embedded green infrastructure into their development model – rooftop gardens, vertical forests, tree-lined expressways. Greenery is more than cosmetic, it's an integral part of policy.

Seoul offers another example. The Seoul Forest, opened in 2005, now contains hundreds of thousands of trees where there was once a water treatment plant. It's a transformation not unlike that of Kibos – land that once served one purpose reclaimed and reimagined as a space for life, for breath. Seoul's initiative was also driven by necessity: to counter air pollution, to offer residents cleaner air and psychological reprieve from dense urban living.

These city forests, though often newer than rural ones, offer a glimpse into a future where urban life and biodiversity are not opposing forces. Trees regulate temperatures, reduce run-off, buffer air pollution and offer critical habitat to birds, insects and small mammals. They also tell stories: of resilience, of resistance, of rootedness.

In Nairobi, the story of Karura Forest is one of all three. Professor Wangari Maathai, whose Green Belt Movement I wrote about in chapter four, led a fierce campaign to save Karura from illegal development in the 1990s. At the time, Karura was seen by developers as vacant land, an opportunity for offices and high-end housing. But for Wangari Maathai and her fellow activists, it was a sacred space. A breathing lung for the city. A sanctuary of indigenous trees, some centuries-old. At the time of writing, Karura is still beloved – a green refuge where Nairobians run, walk, bird-watch, and reconnect with nature – but its future is again being actively contested. Government moves to centralize control and revenues, legal disputes over proposed excisions for road expansion, and tensions surrounding the sidelining of the Friends of Karura Forest have pushed this much-loved urban forest back into a phase of political struggle.

There's a cultural element to urban forests that's often overlooked too. They shape our sense of place. They remind us of what came before and what might come again. In cities, where life is often fast and fractured, trees offer continuity. Their presence is a quiet insistence that not all growth has to be vertical. That not all roots are metaphorical.

What Sheffield, Singapore, Seoul and Nairobi have in common is that someone, at some point, made a decision to prioritize green over grey. Trees in cities don't fall out of the sky – they are fought for, planned, protected. They are an investment in a kind of coexistence that refuses the tired binary between human and nature.

*

In cities, where concrete dulls the senses and the pace of life leaves little room for reflection, forests – whether full-grown or pocket-sized – offer a kind of recalibration. Urban forests have been shown to improve mental health, lower cortisol levels and reduce symptoms of anxiety and depression. But beyond the measurable benefits, they offer something harder to name: a softening. A sense of interconnection that dissolves the illusion of separateness. They remind us that we are porous beings, breathing in what trees breathe out, held in place by roots we cannot see. They disrupt the fantasy of autonomy, of standing apart. And instead, they offer a kind of unhurried non-transactional companionship. Being among trees doesn't ask us to perform or produce. It simply invites us to notice. To pay attention. To slow down enough to feel the world moving through us – not just around us. That shift in perception, subtle as it may be, is where healing begins. The kind of healing that comes from remembering that we were never separate to begin with.

Ecological grief is a term we hear more often now – the mourning that comes from watching the natural world unravel. But I wonder if forests also offer a form of ecological therapy. A place to feel, to process, to reimagine. Whether in Kibos or Karura, Sheffield or Seoul, standing among trees changes how we think. They don't offer solutions. They offer perspective, scale, stillness and, sometimes, a kind of remembering. Perhaps that's what forests have always done. In myth and medicine, in ritual and recovery, they have held a mirror up to us – not just showing us who we are, but who we might become.

*

The therapy that trees provide has never been purely 'natural'. It straddles both nature and culture. We often think the healing comes from spending time in a living forest – their shade, their scent, their stillness. But trees carry such a transformative power that even when they are turned into something else – into paper, into a bench, into the wood of a clarinet – they continue to shape and soothe us. That's the beauty of trees. They don't just grow, they speak. Kilmer's words, in his poem 'Trees', have been quoted so often that they can sound sentimental, but as I stand among trees I hear them as something true.

I think that I shall never see
A poem lovely as a tree.

A tree whose hungry mouth is prest
Against the earth's sweet flowing breast;

A tree that looks at God all day,
And lifts her leafy arms to pray;

A tree that may in summer wear
A nest of robins in her hair;

Upon whose bosom snow has lain;
Who intimately lives with rain.

Poems are made by fools like me,
But only God can make a tree.[5]

*

That impulse, to hear trees, finds echoes in cities everywhere. Some of the most surprising art in urban spaces tries to do exactly that: give trees a voice. In New York, Christopher Janney's *Sonic Forest* transformed the Lincoln Centre into something alive. Twenty-five slender columns, tree-like in shape, responded to footsteps and gestures, answering with cascades of marimba tones, frog calls, percussion: a whole grove of sound, never the same twice. You could stand there in the middle of the city and feel a forest answering you back – not with leaves, but with music. This installation has toured the world, behaving like trees in each of the cities it appears in, enjoining people to gather and also to listen.[6]

*

In Douala, the largest city in Cameroon, Lucas Grandin's *Le Jardin sonore* rises above the edge of the Wouri River like a vertical forest – three tiers of wooden walkways wrapped in plants. It began as a response to a neglected riverside, a place that had become a dumping ground. In 2010, Grandin invited the residents of Bonamouti, a suburb of the city, to imagine something different. Together they cleared the site, planted a vertical garden, and built a wooden tower that now overlooks the mangroves and the wide, tidal sweep of the river. The structure collects rainwater in barrels, stores it for months, and then releases it in slow, deliberate drops. Those drops fall into metal cans of different sizes, each one tuned so that water becomes percussion – a gentle, irregular melody that you can hear as you climb from level to level. At the same time, the drops irrigate the plants, so that every sound is also an act of care.

This is a garden you can listen to. A place where flowers and leaves grow out of sound, where the rhythm of falling water softens the noise of the city around it. On hot afternoons, the upper floors of the structure catch a breeze, and the city opens out below: the river, the mangroves, the tide sliding out to sea.[7]

In Philadelphia, the Intra-Galactic Forest workshops wove together art, sound, ground and community beneath the oak canopy of Awbury Arboretum. Created as part of the S(tree)twork initiative, this multi-year public art project sought to awaken new ways of connecting with trees – not just as background greenery, but as beings whose rhythms, frequencies and presence can be heard and felt.

Residents, artists, sound engineers, ecologists, drummers and printmakers all gathered during September 2021 to explore what it means to slow down and listen deeply to the non-human world. Field recordings of birdsong, water flow and wind were collected by hand, then transformed in workshops using sensors: wind changed playback speed, water flow altered rhythm. The resulting compositions – collaboratively crafted – became part of a performance called *Summoning the Future Forest with the Sun Ra Arkestra*.

That afternoon, more than 650 people reclined on the grass. A procession of horn players, led by Sun Ra's Marshall Allen, approached across the meadow, followed by drums carved from trees felled in the arboretum itself – slit drums made from silent-fall trees, restored to sound and community by local craftspeople and artists. Meanwhile, participants pressed leaves into ink prints, and NASA data of Saturn's plasma waves was translated into shimmering soundscapes echoing through the trees.

The concert was not an advert for tree-planting: it was an invocation, an imagining of a forest not just as ecosystem but as sonic, multispecies companion. As the Sun Ra Arkestra played amid the oaks, it felt almost as though the trees themselves had become instruments, their presence animated by breath and wood and intentional intent. This was not merely performance; it was ecology as art. Music rising from soil, from fallen timber, from leaves still alive. Drumbeats carved from dead wood, carried forward in rhythm to summon a living future forest.[8]

*

These projects make visible something I have always felt: music and trees share a root system. They show how deeply sound comes from wood, and how wood carries memory. The clarinet I picked up not long before the pandemic saved me through months of lockdown. And the day I learned that most clarinets are made of mpingo (*Dalbergia melanoxylon*) – a tree native to East Africa – I felt something close, something circular. It was as if every note I played was drawing breath through a tree that once grew not so far from where I grew up.

My daughter plays the recorders: soprano, alto, tenor, bass. One day she said that playing them feels like an extension of her breath, like the wood is breathing with her. I think about that often. Breath as a bridge. Between forest and body. Between city and soil. Out of that came my project 'Plant an Orchestra', which involves planting 9,000 trees: a thousand in honour of each of the children in the next generation of our family. For shade, for soil, for sound. Perhaps, decades from now, someone will make an instrument out of one of those trees, and another breath will pass through it, making the wood sing again. I try to choose trees with that in mind. Each sapling feels like a note. Each note, part of a score I will never hear in full (the trees will reach maturity after I am dead) but can still begin – one breath, one tree at a time.

I am not alone in thinking this way. Roosevelt Island in New York now has its own 'healing forest', 47 species of native trees planted with the help of scientists, city planners and Lenape elders. They placed small tags on some of the trees, tags that jingle like wind chimes when the breeze moves through. The sound is light, almost tentative, but it is enough to remind anyone walking there that these trees hold more than roots. They hold a way of listening.

Urban forests like this, and like the arboretum where this chapter began, reveal that trees and music are not separate worlds. They grow from the same elements: wood, wind, patience. They show us that a city, too, is a kind of composition. You could stand in the middle of Nairobi and hear it if you wanted: the hiss of leaves, the squish of jacaranda blossoms across pavement, the call of a horn from the street, the rise of a human voice.

I often think the forest's greatest gift to the city isn't just shade. It's sound. The way it teaches us to listen. In the city, sound is how we remember presence, absence, growth. How we reclaim time. We plant for more than soil; we plant for story, for breath, for song.

If a tree can become a clarinet, and a clarinet a voice, then perhaps a city can become a kind of orchestra, made not just of concrete but of what we pay attention to. When I plant trees, I am writing something into the future: a composition of leaves and air. Cities can be like that too – written season by season, note by note, until one day they surprise us with their music.

And so, when I think of the future, I think of cities that sound like forests: places where our presence is less about control and more about listening. Places where wood and breath, stone and shade, work together like parts in a piece of music.

The forests of the future will not emerge solely from policy changes or technological advancements. They will grow from the stories we choose to tell – about what it means to live well, to belong, to care. We need new myths now: not of conquest, but of co-creation. Not of extraction, but of return. Stories where trees are not scenery, but kin. Where forests are not just places, but practices. So rather than seeking to reduce, or eliminate, the human element, perhaps 'rewilding' should instead seek a different kind of presence. The forests that endure are not those kept untouched, but those kept in relationship – across seasons, across generations. We don't need to rewind time to belong in these ecosystems. We've been in them all along.

Epilogue

A Living City: Ever Becoming

I'm walking up to the country park round the corner from where I live. It's just gone 5am and the spring light is only beginning to filter through the branches. The air is crisp. The stillness feels sacred. As the hill steepens, I slow down, which turns out to be the right thing to do. In among the trees, I see a stag. His antlers are proud as we take each other in. I see the shiver run through his flanks. I don't move. Minutes pass. Neither of us shifts. My senses awaken.

Something rustles in the undergrowth. We both startle. An early morning jogger rushes past, leaping over a fallen branch, missing everything. The stag bounds into the trees. Near the top, I walk through the empty car park towards the view of Oxford I love. There, I catch sight of a hare. I keep moving slowly. We both pause. In the foreground, the hare. Down below, the city. We are both facing home.

In that moment I had shared a sense of coexistence with the stag and the hare. I was not alone in the landscape. These are moments that cities rarely prepare us for, yet they reveal something essential: that life moves through our spaces whether we notice it or not. And maybe the first task is to remember to notice nature. Cities are not the opposite of nature. They are a part of it. Cities are often framed as human triumphs over the wild: paved, regulated, insulated from the unpredictable rhythms of the natural world. But that framing has always been flawed. Even beneath the densest concrete, roots push upward. Even through traffic and noise, birds find ways to sing.

Cities, I've come to believe, are not fixed or inert. They are alive.

I long to live as if that were true, to see cities not as machines but as ecologies, messy and layered and full of potential. I want to understand them as breathing and relational instead of efficient and engineered. A living city pulses with memory. It holds the footprints of those who've walked its streets, the dreams of those who've planted seeds in its corners, the scents of kitchens and market stalls and blossom-heavy avenues.

*

I return often to the frogs that came to our garden water features, uninvited but very much expected, somehow. We hadn't curated or designed their arrival. All we had done was offer an invitation – a vessel, a place, some still water. And that was enough. They came. Life came. As though it had simply been waiting for someone to notice.

That idea has taken root in me: that perhaps nature doesn't vanish, but lies dormant, expectant, listening. And when we extend a welcome, however small, it often responds with extraordinary grace. I see it in the unexpected buzz of pollinators on a city verge. In the sprouting of greens on forgotten railway land. In the soft hum of cicadas from a recording on my phone, my anchor to the Kenya of my childhood, and a prayer for the Europe I now live in.

When I think of cities, I think of them as layered invitations. Places where we can choose what kind of relationship we want to cultivate – with each other, with the land, with the more-than-human world that persists around and beneath us. Cities don't have to be spaces of exclusion or erasure. They can be regenerative. Hospitable.

But that kind of aliveness doesn't happen by accident. It happens through practice. Through attention. Through remembering that, just as we shape our environments, they shape us in return. It's ecological, political, emotional, ancestral. It's about the stories we tell ourselves about who belongs, and who or what deserves care.

In Swahili, the word for organic is *uhai* – a word that means 'with life' or 'aliveness'. I've always loved that. It captures something that English sometimes flattens. Uhai is not just the absence of anything unnatural – it's the presence of vitality, of flow. When I think about cities being alive, I think of that layered meaning. A city that works well is not just efficient or populous – it is *with life*. It pulses, it breathes, it welcomes.

For cities to truly thrive, they must move like uhai – not in opposition to nature, but with it. Not in rigid grids alone, but in curving paths, in seasonal rhythms, in ways that allow both humans and more-than-humans to flourish. This is not sentiment. It's structure. It's design.

And it's a choice.

*

I make my way back down from the park, the light lifting now, and stumble into a more modern rhythm of the city: a young woman on an e-scooter, heading to work. She slows down, leaning slightly towards me as she whispers, almost conspiratorially, and points at the lane I'm about to turn into.

'There's a badger,' she says.

I smile, already imagining the sway of Bumble's lolloping walk.

'Wonderful,' I say.

She beams back – 'Wonderful!' – and pushes off, disappearing down the hill, her voice a wave goodbye.

I keep walking, light-footed now, thinking of how these brief, unlikely moments – stag, hare and badger – thread themselves through the heart of a city morning. They are the reminders I carry: that life is always here, even in the most ordinary corners, waiting for us to notice.

Notes

CHAPTER ONE. BORDER CROSSERS: URBAN ANIMALS AND THE BOUNDARIES THEY BLUR

1. J Jowit (2004), 'We're born to be wild'. Available: https://www.theguardian.com/uk/2004/apr/18/highereducation.research [accessed 9 March 2025].
2. M Peplow (2004), 'Lab rats go wild in Oxfordshire', *Nature*.
3. A documentary about this, *The Laboratory Rat: A Natural History* (dir. Manuel Berdoy) can be found at https://www.youtube.com/@ratlifeorg
4. Homer (1898), trans. S. Butler, *The Iliad of Homer: rendered into English prose for the use of those who cannot read the original*, London, New York: Longman's, Green.
5. L Owen, (2013) 'Sminthean Apollo'. Available: https://mouseinterrupted.wordpress.com/2013/01/13/sminthean-apollo/ [accessed 22 December 2024].
6. P L Shipman (2014), 'The bright side of the Black Death', *American Scientist*, 102(6): 410–413.
7. K R Dean, F Krauer, L Walløe, O C Lingjærde, B Bramanti, N C Stenseth, and B V Schmid (2018), 'Human ectoparasites and the spread of plague in Europe during the second pandemic', *PNAS*, 115(6): 1304–1309.
8. It's important to note that disease transmission is bidirectional – humans can contract diseases from animals, and we can also infect them. This dynamic

has been a constant throughout history. The pace, however, has accelerated in the Anthropocene due to significant human impact on the environment. As we encroach on natural habitats and adapt to urban life, we find ourselves living in closer proximity to wildlife, which further increases the speed of disease transmission. Poor sanitation exacerbates the problem. Ultimately, because animals are drawn to cities by the availability of food and, to some extent, companionship, they will continue to share these spaces with us. It is essential that we focus on understanding how to coexist with them more effectively.

9. R H B K Clapperton and New Zealand Department of Conservation (2006), *A Review of the Current Knowledge of Rodent Behaviour in Relation to Control Device*s, Science & Technical Pub., Department of Conservation.
10. R Sullivan (2004), *Rats: Observations on the History and Habitat of the City's Most Unwanted Inhabitants,* New York: Bloomsbury.
11. A Braun (2013), 'Her Majesty's Rat-Catcher'. Available: https://www.laphamsquarterly.org/roundtable/her-majestys-rat-catcher [accessed 22 December 2024].
12. W Janusczak (2005), 'Art: Who's afraid of the big bad guy?' Available: https://waldemar.tv/2005/10/art-whos-afraid-of-the-big-bad-guy/ [accessed 22 December 2024].
13. M P de Cock, H J Esser, W H M Van der Poel, H Sprong and M Maas (2024), 'Higher rat abundance in greener urban areas', *Urban Ecosystems*, 27: 1389–1401.
14. J T Hagstrum (2013), 'Atmospheric propagation modeling indicates homing pigeons use loft-specific infrasonic "map" cues', *Journal of Experimental Biology*, 216(4): 687–699.
15. C Homen (2015), *Cher ami*, University of Central Oklahoma/ProQuest Dissertations & Theses Global. Available: https://www.proquest.com/docview/1755696645 [accessed 22 December 2024].
16. *Economist* (2001), 'Road rage', *Economist*. Available: https://www.economist.com/britain/2001/03/08/road-rage [accessed 22 December 2024].

CHAPTER TWO. EDGES AND ENCLOSURES: HOW GARDENS HELP US RETHINK BELONGING

1. C P Y Lim, D Dillon and P K H Chew (2020), 'A guide to nature immersion: psychological and physiological benefits', *International Journal of Environmental Research and Public Health*, 17(16).

2. Guerrilla gardening means gardening without permission, typically on land that is neglected, abandoned or otherwise not open to public cultivation. It is often a form of direct action – part protest, part planting – used to reclaim space, beautify urban environments, and challenge assumptions about who controls land and how it is used.
3. P Linebaugh (2008), *The Magna Carta Manifesto: The Struggle to Reclaim Liberties and Commons for All*, Berkeley: University of California Press.
4. Anonymous, 'The Goose and the Common' (circa 1700s), a traditional English protest poem against the enclosure of common land, first recorded in full in 1821 in *The Tickler Magazine*.
5. J M Neeson (1993), *Commoners: Common Right, Enclosure and Social Change in England, 1700–1820*, Cambridge: Cambridge University Press.
6. E P Thompson (2013), *The Making of the English Working Class*, London: Penguin Books.
7. Open Spaces Society (2020), 'Lancaster green space saved for the community'. Available: https://www.oss.org.uk/lancaster-green-space-saved-for-the-community/ [accessed 14 September 2025]
8. E Baigent (2016), 'Octavia Hill, nature and open space: crowning success or campaigning "utterly without result"', in E Baigent and B Cowell (eds), *Octavia Hill, Social Activism and the Remaking of British Society*, London: University of London Press.
9. O Hill and C E Maurice (1914), *Life of Octavia Hill as Told in her Letters*, London: Macmillan.
10. Hextable Heritage Society (2003), 'The Remarkable Women of Swanley Horticultural College: Seven Short Stories'. Available: https://www.hextable-heritage.co.uk/Remarkable%20women.pdf
    L Broadbent (2023), 'Women Who Meant Business'. Available: https://womenwhomeantbusiness.com/2023/05/30/fanny-wilkinson-1855-1951/ [accessed 5 May 2025].
11. MPGA (2022), 'Blazing a Trail'. Available: https://www.mpga.org.uk/pdfs/Insight_GDJJuly2022_copyrightSocietyofGardenDesigners.pdf [accessed 5 May 2025].
12. P Johnson (2012), 'The Eden Project – gardens, utopia and heterotopia'. Available: https://silo.tips/download/the-eden-project-gardens-utopia-and-heterotopia [accessed 4 March 2021].
13. Broadbent (2023), 'Women Who Meant Business'.

14. Hill and Maurice, *Life of Octavia Hill as Told in her Letters*.
15. The creation of Red Cross Garden was also a showcase of female-forward collaboration. Hill may have spearheaded the project, but she had crucial help from other women. Funding came largely from Julia, the Countess of Ducie, a philanthropist who trusted Hill's vision. The garden's layout was planned with the aid of Fanny Wilkinson, then the MPGA's resident landscape designer. In fact, Red Cross Garden became one of MPGA's proud early achievements, uniting charitable donors, reformers and the local community. Another woman, Emmeline Sieveking (a member of Hill's circle and the Kyrle Society), assisted on the ground with implementation. Their combined talents produced a space of striking beauty and utility.
16. T Brown (2013). 'The making of urban "healtheries": the transformation of cemeteries and burial grounds in late-Victorian East London.' *Journal of Historical Geography*, 42: 12–23.
17. Barnett, a social activist and wife of the clergyman Samuel Barnett, spent the 1880s and 1890s working among the poor of Whitechapel (she co-founded the educational Toynbee Hall and the Whitechapel Art Gallery). Like Hill, she saw how crushing poverty was aggravated by the ugliness and monotony of slum surroundings.
18. Hampstead Garden Suburb Heritage (2015). Available: https://www.hgsheritage.org.uk/Detail/collections/H120 [accessed 5 May 2025].
19. Housing density would be capped at eight houses per acre (an incredibly low density for London), ensuring plenty of open garden space around each home. Every house, whether a grand villa or a small cottage, was to have a garden or be adjacent to communal greens. Roads were to be lined with trees and kept wide, and, notably, no walls would separate plots, and hedges and flower fences would create a feeling of openness and green continuity throughout the estate. The suburb reserved large public parks and woodlands accessible to all residents 'without regard to the amount of their rent', so that the best landscapes were truly shared, not locked behind private gates. Even details like forbidding loud church bells were considered, to maintain tranquillity. In essence, Barnett took the ethos of the Victorian women's garden movement – beauty for all, healthful open air and communal benefit – and baked it into an urban planning blueprint.

NOTES

CHAPTER THREE. FEAR AS COEXISTENCE: POWER IN THE SAME SPACE

1. Apollodorus, trans. R Hard (1998), *The Library of Greek Mythology*. Oxford: Oxford University Press.
2. 'The Nemean Lion', Perseus digital library, Tufts University. Available: https://www.perseus.tufts.edu/Herakles/lion.html [accessed 7 October 2024].
3. Discovery (2022), 'The truth about the European lion'. Available: https://www.discoveryuk.com/big-cats/the-truth-about-the-european-lion/ [accessed 1 November 2024].
4. Novinite.com (2024), 'Archaeological marvel: lions hunted in Bulgaria 5,000 years ago'. Available: https://www.novinite.com/articles/223698/Archaeological+Marvel%3A+Lions+Hunted+in+Bulgaria+5%2C000+Years+Ago [accessed 1 November 2024].
5. R Turere (2013), 'My invention that made peace with lions', TED. Available: https://www.ted.com/talks/richard_turere_my_invention_that_made_peace_with_lions [accessed 1 November 2024].
6. L R Waller (1976), 'The Maasai and the British 1895–1905: The origins of an alliance', *The Journal of African History*, 17(4): 529–553.
7. T Spear and R Waller (eds) (1993), *Being Maasai: Ethnicity and Identity in East Africa*, Martlesham: Boydell & Brewer.
8. NPR (2016), 'What happened when the lions got loose in Nairobi'. Available: https://www.npr.org/sections/thetwo-way/2016/02/19/467338157/what-happened-when-the-lions-got-loose-in-nairobi [accessed 24 September 2025].
9. The World (2016), 'Lions stalk Nairobi's suburbs'. Available: https://theworld.org/stories/2016/07/31/lions-stalk-nairobi-s-suburbs [accessed 24 September 2025].
10. Kenya, thankfully, banned trophy hunting in 1977. This is not the case in Tanzania, with which Kenya shares a 777km (about 483 miles) border that animals cross over, back and forth.
11. E W Ashe (1808), 'Letter from E. Wellesley Ashe to Theodore Roosevelt', in Theodore Roosevelt Papers, Theodore Roosevelt Digital Library, Dickinson State University.
12. Smithsonian Institute (1909), 'Roosevelt African Expedition Collects for SI'. Available: https://siris-sihistory.si.edu/ipac20/ipac.jsp [accessed 6 December 2024].

13. T Roosevelt (1910), *African Game Trails*, New York: Charles Scribner's Sons.
14. J Waithaka (2012), 'Historical factors that shaped wildlife conservation in Kenya', *The George Wright Forum*, 29(1): 21–29.
15. M Cowie (1963), *I Walk with Lions: The Story of Africa's Great Animal Preserves, the Royal National Parks of Kenya, as told by their First Director*, New York: Macmillan [1963, c.1961].
16. M Venkataraman, P J Johnson, A Zimmermann, R A Montgomery and D W Macdonald (2021), 'Evaluation of human attitudes and factors conducive to promoting human–lion coexistence in the Greater Gir landscape, India', *Oryx*, 55(4): 589–598.
17. D Spranger (2023), 'U-M-led study investigates lions' interactions with humans in a diminishing habitat', Michigan News, University of Michigan. Available: https://news.umich.edu/u-m-led-study-investigates-lions-interactions-with-humans-in-a-diminishing-habitat/ [accessed 8 December 2024].
18. P Chardonnet, P Mésochina, P -C Renaud, C Bento, D Conjo, A Fusari, C Begg, M Foloma and F Pariela (2009), 'Conservation Status of the Lion (*Panthera leo* Linnaeus, 1758) in Mozambique'. Available: https://agritrop.cirad.fr/550720/1/document_550720.pdf [accessed 8 December 2024].
19. Sandra McPherson, 'Lions', from *Elegies for the Hot Season*, Bloomington: Indiana University Press (1970); https://www.poetryfoundation.org/poems/42857/lions

CHAPTER FOUR. HAWKS, IBISES AND OTHER SKY NEIGHBOURS

1. A verse from my poem 'Brood', which reflects on this episode, and has since been anthologized, first published in 2016 in *The Lamp Literary Journal*, Volume VI, Queen's University, Kingston Ontario.
2. eBird, Cornell University, 'Black Kite, *Milvus migrans*'. Available: https://ebird.org/species/blakit1 [accessed 24 December 2024].
3. Firehawk species include: black kite (*Milvus migrans*), brown falcon (*Falco berigora*), and whistling kite (*Haliastur sphenurus*).
4. 'The Firebird', *Dreamtime Dreams*, BBC World Service Archive Project (1999). Available: https://www.bbc.co.uk/programmes/p0356vfy [accessed 1 October 2025].
5. O Pilipili (2009), *The Standard*, 'Marabou storks, a fortune or nuisance?'.

Available: https://www.standardmedia.co.ke/article/1144005945/marabou-storks-a-fortune-or-nuisance [accessed 30 December 2024].
6. C Ogada (2022), InfoNile, 'Fate unknown: urbanisation dwindles home to Marabou Storks in Nairobi'. Available: https://infonile.org/en/2022/11/urbanisation-dwindles-marabou-storks/ [accessed 30 December 2024].
7. E Frank and A Sudarshan (2024), 'The social costs of keystone species collapse: evidence from the Decline of vultures in India', *The American Economic Review*, 114: 3007–3040.
8. This figure is sourced from Avibase, an internet-based repository that tracks and classifies bird species across the globe by taxonomy and habitat range: https://avibase.bsc-eoc.org/checklist.jsp
9. The seven leaders represented were Wangu wa Makeri, Waiyaki wa Hinga, Mekatilili wa Menza, Masaku Ngei, Nabongo Mumia, Ole Lenana and Gor Mahia Wuod Ogalo Nyakwar Ogalo. Wangari Maathai deliberately chose leaders from diverse Kenyan communities, recognizing the complexity and nuance of their colonial resistance – figures whose legacies have often sparked debate. This selection underscores her inclusive and pluralistic approach to questions of environmental and social justice.

CHAPTER FIVE. GROWN IN THE SOIL: CULTIVATING IN CRISIS

1. H Ohly, S Gentry, R Wigglesworth, A Bethel, R Lovell and R Garside (2016), 'A systematic review of the health and well-being impacts of school gardening: synthesis of quantitative and qualitative evidence', *BMC Public Health* 16, article 286.
2. A E Van Den Berg and M H Custers (2011), 'Gardening promotes neuroendocrine and affective restoration from stress', *Journal of Health Psychology*, 16(1): 3–11.
3. S -O Kim, S Y Son, S., M J Kim, C H Lee and S -A Park (2022), 'Physiological responses of adults during soil-mixing activities based on the rresence of soil microorganisms: a metabolomics approach', *Journal of the American Society for Horticultural Science*, 147(3): 135–144.
4. I Panțiru, A Ronaldson, N Sima, A Dregan, and R Sima (2024), 'The impact of gardening on well-being, mental health, and quality of life: an umbrella review and meta-analysis', *Systematic Reviews* 13, article number 45.

5. M O Oyugi and O A K'akumu (2007), 'Land use management challenges for the city of Nairobi', *Urban Forum*, 18: 94–113.
6. C Eliot (1905), *The East Africa Protectorate*, London: Edward Arnold.
7. Ruth Burrows, 'I Made a Garden for God', https://www.stjames-cathedral.org/PoemoftheWeek/burrows-garden.aspx [accessed 20 November 2025]
8. J Harrison (2025), Allotment Garden Web Site, 'Allotment History – A Brief History of Allotments in the UK'. Available: https://www.allotment-garden.org/allotment-information/allotment-history/ [accessed 6 July 2025].
9. Science and Media Museum (2021), Science Museum Group, 'The history of allotments'. Available: https://www.scienceandmediamuseum.org.uk/objects-and-stories/history-allotments [accessed 6 July 2025].
10. Defence of the Realm (Acquisition of Land) Act 1916, CHAPTER 63 6 and 7 Geo 5. Available at: https://www.legislation.gov.uk/ukpga/Geo5/6-7/63 [accessed 6 July 2025].
11. J C Niala (2020), 'Dig for vitality: UK urban allotments as a health-promoting response to COVID-19', *Cities & Health* 5 (figure 1).
12. O E Butler (1993), *Parable of the Sower*, New York: Four Walls Eight Windows.

CHAPTER SIX. HELD IN THE SOIL: PLANTS, MEMORY AND RESILENCE

1. J McCrae, 'In Flanders Fields' (1918), in *Poems of the Great War* (1998), London: Penguin.
2. D Mogîldea and C Biță-Nicolae (2024), ' green space sustainability', *Urban Science*, 8(159).
3. The Blitz refers to the German bombing campaign against the United Kingdom during the Second World War that lasted from 7 September 1940 to 21 May 1941. The name 'Blitz' is derived from the German word *Blitzkrieg*, meaning 'lightning war', and it describes the sudden and intense bombing raids primarily aimed at British cities, particularly London. The Blitz caused widespread destruction and civilian casualties but also demonstrated the resilience of the British population.
4. The Luftwaffe was the aerial warfare branch of the German military during the Second World War. It was responsible for conducting bombing raids, including those during the Blitz. See J Gardiner (2010), *The Blitz: The British Under Attack*, London: HarperPress.
5. G F Vale (1945), 'Bethnal Green's Ordeal 1939–1945', London: Council of the Metropolitan Borough of Bethnal Green.

6. 'The Garden', episode 1 of 'The Root of the Matter', a podcast from the Wellcome Collection (2022). Available: https://wellcomecollection.org/series/YsQLZxEAACAAWQ4J [accessed 6 May 2025].
7. The struggle is a phrase dense with history, pain, resistance and endurance. In South Africa it refers specifically to the collective fight against apartheid: the decades-long battle waged by ordinary people, students, workers, activists and exiled leaders against a system designed to dehumanize. The images that brought it to the world's attention include burning barricades in townships, underground pamphlets printed in secret, mothers protesting in black, songs of freedom sung through tear gas. It was fought in courts and classrooms, on picket lines and through boycotts, and carried in the bodies and memories of those detained, tortured, exiled and killed. There's also a quiet power in the word itself – it's not 'the victory' or 'the resistance' but *the struggle*. Ongoing. Imperfect. Human. It acknowledges effort, suffering and endurance without guaranteeing an end.
8. This system required Black South Africans to carry identification documents at all times, controlling their movement and effectively turning everyday life into a constant negotiation with state power. Without a valid pass, one could be arrested, detained or forcibly relocated.
9. S Pooley (2018), 'The long and entangled history of humans and invasive introduced plants on South Africa's Cape Peninsula', in *Histories of Bioinvasions in the Mediterranean*, S Pooley and A I Queiroz (eds), Springer International Publishing AG.
10. The term 'Coloured' or 'Cape Coloured' refers to a diverse group of people in South Africa, particularly in the Western Cape, who are of mixed ancestry – often including Indigenous Khoisan, enslaved Southeast Asian and African peoples, and European settlers. Under apartheid, it was imposed as a racial category designed to divide and control populations, positioning Coloured people as neither 'white enough' nor 'Black enough'. Most of the people I met from the community still used it to describe themselves, though others are now turning away from it.
11. W Ellis (2019), 'A Tree Walks through the Forest: Milkwoods and Other Botanical Witnesses', *Catalyst: Feminism, Theory, Technoscience*, 5(2).
12. R Haridy (2025), 'An ancient tree revealed the tale of Earth's magnetic field reversal', New Atlas. Available: https://newatlas.com/environment/ancient-tree-earths-magnetic-field-reversal/ [accessed 6 May 2025].
13. The Grenfell Tower fire occurred on 14 June 2017 in North Kensington, West

London. A fire broke out in the 24-storey residential tower block and rapidly spread due to highly flammable cladding installed on the building's exterior. Seventy-two people lost their lives, and many more were injured or displaced. The disaster exposed deep-rooted inequalities in housing safety, neglect of social housing tenants, and institutional failures, sparking national and international outrage and ongoing demands for justice and accountability.

CHAPTER SEVEN. BEYOND REWILDING: HOW CITY INSECTS TEACH US TO SUSTAIN LIFE

1. A P Møller (2019), 'Parallel declines in abundance of insects and insectivorous birds in Denmark over 22 years', *Ecology and Evolution*, 9(11): 6581–6587.
2. L Ball, A Whitehouse, E Bowen-Jones, M Amor, N Anfield, P Hadaway And P Hetherington, 'The Bugs Matter Citizen Science Survey 2024 Report'. Available: https://cdn.buglife.org.uk/2025/04/Bugs-Matter-2024-Report.pdf [accessed 8 May 2025].
3. C A Hallmann, M Sorg, E Jongejans, H Siepel, N Hofland, H Schwan, W Stenmans, A Müller, H Sumser, T Hörren, D Goulson and H De Kroon (2017), 'More than 75 percent decline over 27 years in total flying insect biomass in protected areas', *PLOS One*, 12; https://doi.org/10.1371/journal.pone.0185809 [accessed 8 May 2025].
4. J A Ewald, G R Potts, N J Aebischer, S J Moreby, C J Wheatley and R A Burrell (2024), 'Fifty years of monitoring changes in the abundance of invertebrates in the cereal ecosystem of the Sussex Downs, England', *Insect Conservation and Diversity*, 17(5): 758–787.
5. H Jactel, J -L Imler, L Lambrechts, A -B Failloux, J D Lebreton, Y Le Maho, J -C Duplessy, P Cossart, and P Grandcolas (2020), 'Insect decline: immediate action is needed', *Comptes Rendus Biologies*, 343: 295–296.
6. E Russell (2001), *War and Nature: Fighting Humans and Insects with Chemicals from World War I to Silent Spring*, Cambridge: Cambridge University Press.
7. B Theunissen (2019), 'The Oostvaardersplassen Fiasco', *Isis*, 110(2): 341–345.
8. C R Clement, W M Denevan, M J Heckenberger, A B Junqueira, E G Neves, W G Teixeira and W I Woods (2015), 'The Domestication of Amazonia before European Conquest', *Proceedings of the Royal Society B*, 282; https://doi.org/10.1098/rspb.2015.0813 [accessed 8 May 2025].

9. I Ziffer (2019), 'Pinecone or date palm male inflorescence – metaphorical pollination in Assyrian art', *Israel Journal of Plant Sciences*, 66: 19–33.
10. C K Sprengel (1793), *Das entdeckte Geheimniss der Natur im Bau und in der Befruchtung der Blumen*, Berlin: Friedrich Vieweg.
11. Oxford City Council (2025) 'What we're doing to enhance biodiversity'. Available: https://www.oxford.gov.uk/biodiversity/enhance-biodiversity [accessed 8 May 2025].
12. R Lofthouse (2021), 'Helping city centre pollinators', University of Oxford, Oxford Alumni. Available: https://www.alumni.ox.ac.uk/article/helping-city-centre-pollinators [accessed 9 September 2025].
13. Z An, Q Chen and J Li (2020), 'Ecological strategies of urban eological parks – A case of Bishan Ang Mo Kio Park and Kallang River in Singapore', *E3S Web of Conferences*, 194(3): 05060.
14. J S Ascher, Z W W Soh, S X Chui, E J Y Soh, B M Ho, J X Q Lee, A R Gajanur, and X R Ong (2022), 'The bees of Singapore (Hymenoptera: Apoidea: Anthophila): First comprehensive country checklist and conservation assessment for a Southeast Asian bee fauna', *Raffles Bulletin of Zoology*, 70: 39–64.
15. K Sloan (2017) 'Re-wilding: Cities by Nature', TNOC. Available: https://www.thenatureofcities.com/2017/04/30/re-wilding-cities-nature/ [accessed 10 May 2025].
16. S Schama (2004), *Landscape and Memory*, London: Harper Perennial.
17. A E Samuelson, R J Gill, M J F Brown and E Leadbeater (2018) 'Lower bumblebee colony reproductive success in agricultural compared with urban environments', *Proceedings of the Royal Society B*, 285; https://doi.org/10.1098/rspb.2018.0807 [accessed 9 September 2025].

    K C R Baldock, M A Goddard, D M Hicks, W E Kunin, N Mitschunas, L M Osgathorpe, S G Potts, K M Robertson, A V Scott, G N Stone, I P Vaughan and J Memmott (2015), 'Where is the UK's pollinator biodiversity? The importance of urban areas for flower-visiting insects', *Proceedings of the Royal Society B*, 282; https://doi.org/10.1098/rspb.2014.2849 [accessed 9 September 2025].
18. S Sumner, G Law and A Cini (2018), 'Why we love bees and hate wasps', *Ecological Entomology*, 43(6): 836–845.
19. A Tversky and D Kahneman (1974), 'Judgment under uncertainty: heuristics and biases. Biases in judgments reveal some heuristics of thinking under uncertainty', *Science (American Association for the Advancement of Science)*, 185: 1124–1131.
20. Sumner, Law and Cini, 'Why we love bees and hate wasps'.

CHAPTER EIGHT. SOFT CIVILIZATIONS: HOW WETLANDS SHAPED URBAN WORLDS

1. Luo peoples are a Nilotic ethnolinguistic group found primarily in Kenya, Uganda, Tanzania, South Sudan, Ethiopia and the Democratic Republic of Congo. Subgroups include the Jo-Luo (Kenya), Acholi and Alur (Uganda/DRC), and Anuak (Ethiopia/South Sudan). Despite regional differences, they share cultural traits such as cattle-keeping, clan-based social organization and strong oral traditions.
2. Wisconsin Wetlands Association (2017), '9 Wetland Monsters from World Folklore', adapted from 'A World of Wetland Monsters', by Tod Highsmith (2013). Available: https://www.wisconsinwetlands.org/updates/9-wetland-monsters/ [accessed 10 May 2025].
3. R Janse van Vuuren (2022), 'Grindylow'. Available: https://www.ronelthemythmaker.com/the-faeries-and-folklore-podcast-by-ronel-grindylow-podcast-faeries-folklore/ [accessed 10 May 2025].
4. J N Postgate (1992), *Early Mesopotamia: Society and Economy at the Dawn of History*, London: Routledge.
5. H J Nissen, trans. E Lutzeier (1988), *The Early History of the Ancient Near East, 9000–2000 B.C.*, Chicago: University of Chicago Press.
6. J Jotheri, M Rokan, A Al-Ghanim, L Rayne, M de Gruchy and R Alabdan (2025), 'Identifying the preserved network of irrigation canals in the Eridu region, southern Mesopotamia', *Antiquity*, 99(405): https://doi.org/10.15184/aqy.2025.19 [accessed 27 July 2025].
7. M Zalewski and A Sztuka-Tulińska (2020), 'Urban river restoration: a sustainable strategy for storm-water management in Lodz, Poland', Climate-ADAPT. Available: https://climate-adapt.eea.europa.eu/en/metadata/case-studies/urban-river-restoration-a-sustainable-strategy-for-storm-water-management-in-lodz-poland [accessed 27 July 2025].
8. A Ramsar Wetland City is a designation awarded by the Ramsar Convention on Wetlands to cities that demonstrate exceptional commitment to the conservation and sustainable use of wetlands within their urban and peri-urban boundaries. This voluntary accreditation recognizes cities that integrate wetland protection into urban planning, promote public awareness and engage communities in wetland management.
9. Global Water Partnership (2024), 'Adapting to climate change impacts through

smart irrigation in Ghar El Melh wetland area, Tunisia'. Available: https://www.gwp.org/en/GWP-Mediterranean/WE-ACT/Programmes-per-theme/Climate-Change-Adaptation/adapting-to-climate-change-impact_ghar-el-melh/ [accessed 11 May 2025].
10. BBC News (2013), 'The birthday of Lord Ganesh celebrated on River Mersey'. Available: https://www.bbc.co.uk/news/uk-england-merseyside-24142164 [accessed 11 May 2025].
11. 'Wetland', episode 4 of 'The Root of the Matter', a podcast from the Wellcome Collection (2022). Available at https://wellcomecollection.org/stories/wetland [accessed 11 May 2025].
12. 'Wetland', episode 4 of 'The Root of the Matter', [Accessed 22 Jan 2026]
13. Ibid.
14. Ibid.
15. Poetry Foundation, https://www.poetryfoundation.org/poems/45527/lines-composed-a-few-miles-above-tintern-abbey-on-revisiting-the-banks-of-the-wye-during-a-tour-july-13-1798 [accessed 11 May 2025].

CHAPTER NINE. THE HOOFPRINT BENEATH THE CITY: LEARNING FROM FOOD SYSTEMS OF THE PAST

1. A Beja-Pereira, G Luikart, P R England, D G Bradley, O C Jann, G Bertorelle, A T Chamberlain, T P Nunes, S Metodiev, N Ferrand and G Erhardt (2003), 'Gene-culture coevolution between cattle milk protein genes and human lactase genes', *Nature Genetics*, 35: 311–313.
2. D Ferguson (2024), 'Cambridge could lose city centre cows if council cuts out-of-hours rescue service', *The Guardian*. Available: https://www.theguardian.com/uk-news/2024/nov/10/cambridge-could-lose-city-centre-cows-if-council-cuts-out-of-hours-rescue-service [accessed 8 July 2025].
3. J Whitaker (2016), 'Farm to Table', Restaurant-ing Through History. Available: https://restaurant-ingthroughhistory.com/tag/dairy-lunches/ [accessed 9 July 2025].
4. R J Moore-Colyer (1976), *The Welsh Cattle Drovers: Agriculture and the Welsh Cattle Trade before and during the Nineteenth Century*, Cardiff: University of Wales Press.
5. P J Atkins (1977), 'London's Intra-Urban Milk Supply, circa 1790–1914', *Transactions of the Institute of British Geographers*, 2: 383–399.
6. L Jackson (2015), *Dirty Old London: The Victorian Fight against Filth*, New Haven: Yale University Press.

7. T Almeroth-Williams (2019, *City of Beasts: How Animals Shaped Georgian London*, Manchester: Manchester University Press.
8. T Baker, D K Bolton and P E Croot (1989), 'Paddington: Economic History', British History Online. Available: https://www.british-history.ac.uk/vch/middx/vol9/pp233-241 [accessed 9 July 2025].
9. S Spellen (2010), 'Building of the Day: 1368 Fulton Street', Brownstoner. Available: https://www.brownstoner.com/architecture/building-of-the-36/ [accessed 9 July 2025].
10. A T Hajdik (2014), 'A "Bovine Glamour Girl": Borden milk, Elsie the Cow, and the convergence of technology, animals, and gender at the 1939 New York World's Fair', *Agricultural History*, 88(4): 470–490.
11. G Cusworth, J Lorimer, J Brice and T Garnett (2022), 'Green rebranding: Regenerative agriculture, future-pasts, and the naturalisation of livestock', *Transactions of the Institute of British Geographers*, 47(4): 1009–1027.

CHAPTER TEN. THE LONG GAME: SEEING THE WOOD AND THE TREES

1. E Oriel (2024), 'Jacaranda trees, place and affect: an analysis of Australian newspaper articles, 1900–2023', *Plant Perspectives*. Available: https://www.whp-journals.co.uk/PP/article/view/993 [accessed 9 July 2025].
2. A Birnie (2002), 'The Nairobi Arboretum, Kenya – reflecting the needs of the community for almost a hundred years', *Botanic Gardens Conservation News*, 3(9): 45–49.
3. A C Van Vollenhoven (2020), 'The cultural historical significance of Pretoria's jacaranda trees', *New Contree*, 85.
4. E Oriel, 'Jacaranda trees, place and affect: an analysis of Australian newspaper articles, 1900–2023'.
5. Joyce Kilmer, 'Trees'; https://poets.org/poem/trees [accessed 9 July 2025].
6. Christopher Janney, *Sonic Forest*, Making Architecture Dance; https://www.janneysound.com/project/8/ [accessed 9 July 2025].
7. *Le Jardin Sonore de Bonamouti*, http://lucas.grandin.free.fr/jardinsonore.html [accessed 9 July 2025].
8. Inter-Galactic Forest; https://streetworkproject.net/intra-galactic-forest/ [accessed 9 July 2025].

Bibliography

ALMEROTH-WILLIAMS, T (2019). *City of Beasts: How Animals Shaped Georgian London*. Manchester: Manchester University Press.

AN, Z, CHEN, Q & LI, J (2020). 'Ecological strategies of urban ecological rarks – a case of Bishan Ang Mo Kio Park and Kallang River in Singapore', *E3S Web of Conferences*, 194(3): 05060.

APOLLODORUS, trans. HARD, R (1998). *The Library of Greek Mythology*. Oxford: Oxford University Press.

ASCHER, J S, SOH, Z W W, CHUI, S X, SOH, E J Y, HO, B M, LEE, J X Q, GAJANUR, A R and ONG, X R (2022). 'The bees of Singapore (Hymenoptera: Apoidea: Anthophila): First comprehensive country checklist and conservation assessment for a Southeast Asian bee fauna', *Raffles Bulletin of Zoology*, 70: 39–64.

ASHE, E W (1808). 'Letter from E Wellesley Ashe to Theodore Roosevelt', in Theodore Roosevelt Papers, Theodore Roosevelt Digital Library, Dickinson State University.

ATKINS, P J (1977). 'London's Intra-Urban Milk Supply, circa 1790–1914', *Transactions of the Institute of British Geographers (1965)*, 2: 383–399.

BAIGENT, E (2016). 'Octavia Hill, nature and open space: crowning success or campaigning "utterly without result"', in E Baigent and B Cowell (eds), *Octavia Hill, Social Activism and the Remaking of British Society*. London: University of London Press.

BAKER, T, BOLTON, D K and CROOT, P E (1989). 'Paddington: Economic History', British History Online. Available: https://www.british-history.ac.uk/vch/middx/vol9/pp233-241 [accessed 9 July 2025].

BALDOCK, K C R, GODDARD, M A, HICKS, D M, KUNIN, W E, MITSCHUNAS, N, OSGATHORPE, L M, POTTS, S G, ROBERTSON, K M, SCOTT, A V, STONE, G N, VAUGHAN, I P and MEMMOTT, J (2015). 'Where is the UK's pollinator biodiversity? The importance of urban areas for flower-visiting insects', *Proceedings of the Royal Society B*, 282; https://doi.org/10.1098/rspb.2014.2849 [accessed 9 September 2025].

BALL, L, WHITEHOUSE, A, BOWEN-JONES, E, AMOR, M, BANFIELD, N, HADAWAY, P and HETHERINGTON, P. 'The Bugs Matter Citizen Science Survey 2024 Report'. Available: https://cdn.buglife.org.uk/2025/04/Bugs-Matter-2024-Report.pdf [accessed 8 May 2025].

BBC News (2013). 'The birthday of Lord Ganesh celebrated on River Mersey'.

Available: https://www.bbc.co.uk/news/uk-england-merseyside-24142164 [accessed 11 May 2025].

BEJA-PEREIRA, A, LUIKART, G, ENGLAND, P R, BRADLEY, D G, JANN, O C, BERTORELLE, G, CHAMBERLAIN, A T, NUNES, T P, METODIEV, S, FERRAND, N and ERHARDT, G (2003). 'Gene-culture coevolution between cattle milk protein genes and human lactase genes', *Nature Genetics*, 35: 311–313.

BIRNIE, A (2002). 'The Nairobi Arboretum, Kenya – reflecting the needs of the community for almost a hundred years', *Botanic Gardens Conservation News*, 3(9): 45–49.

BRAUN, A (2013). 'Her Majesty's Rat-Catcher'. Available: https://www.laphamsquarterly.org/roundtable/her-majestys-rat-catcher [accessed 22nd December 2024].

BROADBENT, L (2023). 'Women Who Meant Business'. Available: https://womenwhomeantbusiness.com/2023/05/30/fanny-wilkinson-1855-1951/ [accessed 5 May 2025].

CHARDONNET, P, MESOCHINA, P, RENAUD, P-C, BENTO, C, CONJO, D, FUSARI, A, BEGG, C, FOLOMA, M and PARIELA, F (2009). 'Conservation Status of the Lion (*Panthera leo* Linnaeus, 1758) in Mozambique'. Available: https://agritrop.cirad.fr/550720/1/document_550720.pdf [accessed 8 December 2024].

CLAPPERTON, B K and NEW ZEALAND DEPARTMENT OF CONSERVATION (2006). *A Review of the Current Knowledge of Rodent Behaviour in Relation to Control Devices*. Science & Technical Pub., Department of Conservation.

CLEMENT, C R, DENEVAN, W M, HECKENBERGER, M J, JUNQUEIRA, A B, NEVES, E G, TEIXEIRA, W G and WOODS, W I (2015). 'The domestication of Amazonia before European conquest', *Proceedings of the Royal Society B*, 282; https://doi.org/10.1098/rspb.2015.0813 [accessed 8 May 2025].

eBIRD, CORNELL UNIVERSITY. 'Black Kite, *Milvus migrans*'. Available: https://ebird.org/species/blakit1 [accessed 24 December 2024].

COWIE, M (1963). *I Walk with Lions: The story of Africa's Great Animal Preserves, the Royal National Parks of Kenya, as told by their First Director*. New York: Macmillan [1963, c.1961].

CUSWORTH, G, LORIMER, J, BRICE, J and GARNETT, T (2022). 'Green rebranding: Regenerative agriculture, future-pasts, and the naturalisation of livestock', *Transactions of the Institute of British Geographers*, 47(4): 1009–1027.

DE COCK, M P, ESSER, H J, VAN DER POEL, W H M, SPRONG, H and MAAS, M (2024). 'Higher rat abundance in greener urban areas', *Urban Ecosystems*, 27: 1389–1401.

DEAN, K R, KRAUER, F, WALLØE, L, LINGJÆRDE, O C, BRAMANTI, B, STENSETH, N C, and SCHMID, B V (2018). 'Human ectoparasites and the sread of plague in Europe during the second pandemic', *PNAS*, 115(6): 1304–1309.

DISCOVERY (2022). 'The Truth about the European Lion'. Available: https:// https://www.discoveryuk.com/big-cats/the-truth-about-the-european-lion/ [accessed 1 November 2024].

ECONOMIST (2001). Economist (2001), 'Road rage', Economist. Available: https://www.economist.com/britain/2001/03/08/road-rage [accessed 22 December 2024].

ELIOT, C (1905). *The East Africa Protectorate*. London: Edward Arnold.

ELLIS, W (2019). 'A Tree Walks through the Forest: Milkwoods and Other Botanical Witnesses', *Catalyst: Feminism, Theory, Technoscience*, 5(2).

EWALD, J A, POTTS, G R, AEBISCHER, N, MOREBY, S J, WHEATLEY, C J, and BURRELL, R A (2024). 'Fifty years of monitoring changes in the abundance of invertebrates in the cereal ecosystem of the Sussex Downs, England', *Insect Conservation and Diversity*, 17(5): 758–787.

FERGUSON, D (2024) 'Cambridge could lose city centre cows if council cuts out-of-hours rescue service', *The Guardian*. Available: https://www.theguardian.com/uk-news/2024/nov/10/cambridge-could-lose-city-centre-cows-if-council-cuts-out-of-hours-rescue-service [accessed 8 July 2025].

'THE FIREBIRD', *Dreamtime Dreams*, BBC World Service Archive Project (1999). Available: https://www.bbc.co.uk/programmes/p0356vfy [accessed 1 October 2025].

FRANK, E and SUDARSHAN, A (2024). 'The social costs of keystone species collapse: evidence from the decline of vultures in India', *The American Economic Review*, 114: 3007–3040.

GARDINER, J (2010). *The Blitz: The British Under Attack*. London: HarperPress.

GLOBAL WATER PARTNERSHIP (2024). 'Adapting to Climate Change Impacts through Smart Irrigation in Ghar El Melh wetland area, Tunisia'. Available: https://www.gwp.org/en/GWP-Mediterranean/WE-ACT/

Programmes-per-theme/Climate-Change-Adaptation/adapting-to-climate-change-impact_ghar-el-melh/ [accessed 11 May 2025].

HAGSTRUM, J T (2013). 'Atmospheric propagation modeling indicates homing pigeons use loft-specific infrasonic "map" cues', *Journal of Experimental Biology*, 216(4): 687–699.

HAJDIK, A T (2014). 'A "Bovine Glamour Girl": Borden milk, Elsie the Cow, and the convergence of technology, animals, and gender at the 1939 New York World's Fair', *Agricultural History*, 88(4): 470–490.

HALLMANN, C A, SORG, M, JONGEJANS, E, SIEPEL, H, HOFLAND, N, SCHWAN, H, STENMANS, W, MÜLLER, A, SUMSER, H, HÖRREN, T, GOULSON, D and DE KROON, H (2017). 'More than 75 percent decline over 27 years in total flying insect biomass in protected areas', *PLOS One*, 12; https://doi.org/10.1371/journal.pone.0185809 [accessed 8 May 2025].

HAMSTEAD GARDEN SUBURB HERITAGE (2015). Available: https://www.hgsheritage.org.uk/Detail/collections/H120 [accessed 5 May 2025].

HARIDY, R (2025). 'An ancient tree revealed the tale of Earth's magnetic field reversal', New Atlas. Available: https://newatlas.com/environment/ancient-tree-earths-magnetic-field-reversal/ [accessed 6 May 2025].

HARRISON, J (2025). Allotment Garden Web Site, 'Allotment History – A Brief History of Allotments in the UK'. Available: https://www.allotment-garden.org/allotment-information/allotment-history/ [accessed 6 July 2025].

HEXTABLE HERITAGE SOCIETY (2003). 'The Remarkable Women Of Swanley Horticultural College: Seven Short Stories'. Available: https://www.hextable-heritage.co.uk/Remarkable%20women.pdf [accessed 5 May 2025].

HILL, O, and MAURICE, C E (1914). *Life of Octavia Hill as Told in her Letters*. London: Macmillan.

HOMEN, C (2015). 'Cher Ami'. University of Central Oklahoma/ProQuest Dissertations & Theses Global. Available: https://www.proquest.com/docview/1755696645 [accessed 22 December 2024].

HOMER (1898), trans. BUTLER, S. *The Iliad of Homer: rendered into English prose for the use of those who cannot read the original*. London, New York: Longman's, Green.

HUGHES, L (2006). *Moving the Maasai: A Colonial Misadventure*. Basingstoke: Palgrave, Macmillan.

JACKSON, L (2015). *Dirty Old London: The Victorian Fight against Filth*. New Haven: Yale University Press.

JACTEL, H, IMLER, J-L, LAMBRECHTS, L, FAILLOUX, A-B, LEBRETON,

JD, LE MAHO, Y, DUPLESSY, J-C, COSSART, P, and GRANDCOLAS, P (2020). 'Insect decline: immediate action is needed', *Comptes Rendus Biologies*, 343(3): 295–296.

JANUSCZAK, W (2005). 'Art: Who's afraid of the big bad guy?' Available: https://waldemar.tv/2005/10/art-whos-afraid-of-the-big-bad-guy/ [accessed 22 December 2024].

JANSE VAN VUUREN, R (2025). 'Grindylow'. Available: https://www.ronelthemythmaker.com/the-faeries-and-folklore-podcast-by-ronel-grindylow-podcast-faeries-folklore/ [accessed 10 May 2025].

JOHNSON, P (2012). 'The Eden Project – gardens, utopia and heterotopia'. Available: https://silo.tips/download/the-eden-project-gardens-utopia-and-heterotopia [accessed 4 March 2021].

JOTHERI, J, ROKAN, M, AL-GHANIM, A, RAYNE, L, DE GRUCHY, M, and ALABDAN, R (2025). I'dentifying the preserved network of irrigation canals in the Eridu region, southern Mesopotamia', *Antiquity* 99(405).

JOWIT, J (2004). 'We're born to be wild'. Available: https://www.theguardian.com/uk/2004/apr/18/highereducation.research [accessed 9 March 2025].

KIM, S-O, SON, S, KIM, M, LEE, C, and PARK, S-A (2022). 'Physiological responses of Adults during soil-mixing activities based on the presence of soil microorganisms: a metabolomics approach', *Journal of the American Society for Horticultural Science*, 147(3): 135–144.

LIM, P Y, DILLON, D, and CHEW, P K H (2020). 'A guide to nature immersion: psychological and physiological benefits', *International Journal of Environmental Research and Public Health*, 17(16).

LINEBAUGH, P (2008). *The Magna Carta Manifesto: The Struggle to Reclaim Liberties and Commons for All*. Berkeley: University of California Press.

LIU, C, HERRUP, K, GOTO, S and SHI, B E (2020). 'Viewing garden scenes: Interaction between gaze behavior and physiological responses', *Journal of Eye Movement Research*, 13(1).

LOFTHOUSE, R (2021). 'Helping city centre pollinators', University of Oxford, Oxford Alumni. Available: https://www.alumni.ox.ac.uk/article/helping-city-centre-pollinators [accessed 9 September 2025].

MCCRAE, J (1918). 'In Flanders Fields', in *Poems of the Great War* (1998). London: Penguin.

MOGÎLDEA, D, and BIȚĂ-NICOLAE, C (2024). 'Ruderal plant diversity as a driver for urban green space sustainability', *Urban Science*, 8(159).

MØLLER, A P (2019). 'Parallel declines in abundance of insects and insectivorous birds in Denmark over 22 years', *Ecology and Evolution*, 9(11): 6581–6587.

MOORE-COLYER, R J (1976). *The Welsh Cattle Drovers: Agriculture and the Welsh Cattle Trade before and during the Nineteenth Century*. Cardiff: University of Wales Press.

MPGA (2022). 'Blazing a Trail'. Available: https://www.mpga.org.uk/pdfs/Insight_GDJJuly2022_copyrightSocietyofGardenDesigners.pdf [accessed 5 May 2025].

NEESON, J M (1993). *Commoners: Common Right, Enclosure and Social Change in England, 1700–1820*. Cambridge: Cambridge University Press.

'THE NEMEAN LION', Perseus digital library, Tufts University. Available: https://www.perseus.tufts.edu/Herakles/lion.html [accessed 7 October 2024].

NIALA, J C (2020). 'Dig for vitality: UK urban allotments as a health-promoting response to COVID-19', *Cities & Health* 5 (figure 1).

NISSEN, H J, trans. LUTZEIER, E (1988). *The Early History of the Ancient Near East, 9000–2000 B.C.* Chicago: University of Chicago Press.

NOVINITE.COM (2024). 'Archaeological Marvel: Lions Hunted in Bulgaria 5,000 Years Ago'. Available: https://www.novinite.com/articles/223698/Archaeological+Marvel%3A+Lions+Hunted+in+Bulgaria+5%2C000+Years+Ago [accessed 1 November 2024].

OPEN SPACES SOCIETY (2020), 'Lancaster green space saved for the community'. Available: https://www.oss.org.uk/lancaster-green-space-saved-for-the-community/ [accessed 14 September 2025]

OXFORD CITY COUNCIL (2025). 'What we're doing to enhance biodiversity'. Available: https://www.oxford.gov.uk/biodiversity/enhance-biodiversity [accessed 8 May 2025].

OGADA, C (2022). 'Fate unknown: urbanisation dwindles home to Marabou Storks in Nairobi'. Available: https://infonile.org/en/2022/11/urbanisation-dwindles-marabou-storks/ [accessed 30 December 2024].

OHLY, H, GENTRY, S, WIGGLESWORTH, R, BETHEL, A, LOVELL, R, and GARSIDE, R (2016). 'A systematic review of the health and well-being impacts of school gardening: synthesis of quantitative and qualitative evidence', *BMC Public Health* 16, article 286.

ORIEL, E (2024). 'Jacaranda trees, place and affect: an analysis of Australian newspaper articles, 1900–2023', *Plant Perspectives*. Available: https://whp-journals.co.uk/PP/article/view/993 [accessed 9 July 2025].

OWEN, L (2013). 'Sminthean Apollo'. Available: https://mouseinterrupted.wordpress.com/2013/01/13/sminthean-apollo/ [accessed 22 December 2024].

OYUGI, M O, and K'AKUMU, O A (2007). 'Land use management challenges for the city of Nairobi', *Urban Forum*, 18: 94–113.

PANȚIRU, I, RONALDSON, A, SIMA, N, DREGAN, A and SIMA, R (2024). 'The impact of gardening on well-being, mental health, and quality of life: an umbrella review and meta-analysis', *Systematic Reviews* 13, article number 45.

PEPLOW, M (2004). 'Lab rats go wild in Oxfordshire', *Nature*.

PILIPILI, O (2009). 'Marabou storks, a fortune or nuisance?'. Available: https://www.standardmedia.co.ke/article/1144005945/marabou-storks-a-fortune-or-nuisance [accessed 30 December 2024].

POOLEY, S (2018). 'The long and entangled history of humans and invasive introduced plants on South Africa's Cape Peninsula', in *Histories of Bioinvasions in the Mediterranean*, S Pooley and A I Queiroz (eds). Springer International Publishing AG.

POSTGATE, J N (1992). *Early Mesopotamia: Society and Economy at the Dawn of History*. London: Routledge.

ROOSEVELT, T (1910). *African Game Trails*. New York: Charles Scribner's sons.

RUSSELL, E (2001). *War and Nature: Fighting Humans and Insects with Cchemicals from World War I to Silent Spring*. Cambridge: Cambridge University Press.

SAMUELSON, A E, GILL, R J, BROWN, M J F, LEADBEATER, E (2018). 'Lower bumblebee colony reproductive success in agricultural compared with urban environments', *Proceedings of the Royal Society Publishing B*, 285; https://doi.org/10.1098/rspb.2018.0807 [accessed 9 September 2025].

SCHAMA, S (2004). *Landscape and Memory*. London: Harper Perennial.

SHIPMAN, P L (2014). 'The bright side of the Black Death', *American Scientist*, 102(6): 410–413.

SLOAN, K (2017). 'Re-wilding: Cities by Nature'. Available: https://www.thenatureofcities.com/2017/04/30/re-wilding-cities-nature/ [accessed 10 May 2025].

SCIENCE AND MEDIA MUSEUM (2021), Science Media Group, 'The history of allotments'. Available: https://www.scienceandmediamuseum.org.uk/objects-and-stories/history-allotments [accessed 6 July 2025].

SMITHSONIAN INSTITUTE (1909). 'Roosevelt African Expedition Collects for SI'. Available: https://siris-sihistory.si.edu/ipac20/ipac.jsp [accessed 6 December 2024].

SPEAR, T, and WALLER, R (eds) (1993). *Being Maasai: Ethnicity and Identity in East Africa*. Martlesham: Boydell & Brewer.

SPELLEN, S (2010). 'Building of the Day: 1368 Fulton Street', Brownstoner. Available: https://www.brownstoner.com/architecture/building-of-the-36/ [accessed 9 July 2025].

SPRANGER, D (2023). 'U-M-led study investigates lions' interactions with humans in a diminishing habitat', Michigan News, University of Michigan. Available: https://news.umich.edu/u-m-led-study-investigates-lions-interactions-with-humans-in-a-diminishing-habitat/ [accessed 8 December 2024].

SPRENGEL, C K (1793). *Das entdeckte Geheimniss der Natur im Bau und in der Befruchtung der Blumen*. Berlin: Friedrich Vieweg.

SULLIVAN, R (2004). *Rats: Observations on the History and Habitat of the city's most unwanted inhabitants*. New York: Bloomsbury.

SUMNER, S, LAW, G, and CINI, A (2018). 'Why we love bees and hate wasps', *Ecological Entomology*, 43(6): 836–845.

TAYLOR, R H, and THOMAS, B W (1989). 'Eradication of Norway rats (*Rattus Norvegicus*) from Hawea Island, fiordland, using brodifacoum', *New Zealand Journal of Ecology*, 12: 23–32.

THEUNISSEN, B (2019). 'The Oostvaardersplassen Fiasco', *Isis*, 110(2): 341–345.

THOMPSON, E P (2013). *The Making of the English Working Class*. London: Penguin Books.

TURERE, R (2013). 'My invention that made peace with lions'. TED. Available: https://www.ted.com/talks/richard_turere_my_invention_that_made_peace_with_lions [accessed 1 November 2024].

TVERSKY, A, and KAHNEMAN, D (1974). 'Judgment under uncertainty: heuristics and biases. Biases in judgments reveal some heuristics of thinking under uncertainty', *Science (American Association for the Advancement of Science)*, 185: 1124–1131.

VALE, G F (1945). 'Bethnal Green's Ordeal 1939–1945'. London: Council of the Metropolitan Borough of Bethnal Green.

VAN DEN BERG, A E, and CUSTERS, M H (2011). 'Gardening promotes neuroendocrine and affective restoration from stress', *Journal of Health Psychology*, 16(1): 3–11.

VAN VOLLENHOVEN, A C (2020). 'The cultural historical significance of Pretoria's jacaranda trees', *New Contree*, 85.

VENKATARAMAN, M, JOHNSON, P J, ZIMMERMANN, A,

BIBLIOGRAPHY

MONTGOMERY, R A and MACDONALD, D W (2021). 'Evaluation of human attitudes and factors conducive to promoting human-lion coexistence in the Greater Gir landscape, India', *Oryx*, 55(4): 589-598.

WAITHAKA, J (2012). 'Historical factors that shaped wildlife conservation in Kenya', *The George Wright Forum*, 29(1): 21-29.

WALLER, R (1976). 'The Maasai and the British 1895-1905: The Origins of an Alliance', *The Journal of African History*, 17(4): 529-553.

WHITAKER, J (2016). 'Farm to Table', Restaurant-ing Through History. Available: https://restaurant-ingthroughhistory.com/tag/dairy-lunches/ [accessed 9 July 2025].

WISCONSIN WETLANDS ASSOCIATION (2017). '9 Wetland Monsters from World Folklore', adapted from 'A World of Wetland Monsters', by Tod Highsmith (2013). Available: https://www.wisconsinwetlands.org/updates/9-wetland-monsters/ [accessed 10 May 2025].

ZALEWSKI, M, and SZTUKA-TULIŃSKA, A (2020). 'Urban river restoration: a sustainable strategy for storm-water management in Lodz, Poland', Climate-ADAPT. Available: https://climate-adapt.eea.europa.eu/en/metadata/case-studies/urban-river-restoration-a-sustainable-strategy-for-storm-water-management-in-lodz-poland [accessed 27 July 2025].

ZIFFER, I (2019). 'Pinecone or date palm male inflorescence – metaphorical pollination in Assyrian art', *Israel Journal of Plant Sciences*, 66: 19-33.

Index

Aboriginal folklore *see* Australian Aboriginal stories
Achillea millefolium 110
Adams Event in human history 118
Aden, Uncle 11–12
Aesop's tale of Androcles and the lion 52–3
agency 93–4
agriculture
   and the flooding of the Nile 157
   high-tech industrial farming 181
   monocultures 125, 181
   organic farm in Kibera 93–8
   pesticides and insect decline 125–6
   pollination in 132
   regenerative 181
   rewilding 127–31
   wetlands 156, 160
   see also allotments
air pollution 195
All That Breathes documentary 73

Allen, Marshall 200
Allotment Holder's Enemies 102
allotments 98–103, 104, 121, 137
   the 1918 Allotment project 100–3, 107, 126
   and pollinating insects 132, 133, 138
Alwin (ecologist) 123–4, 126
Amazon rainforest 130, 137
An American Tail 17
Ancient Greece
   lions in 52–6
antelopes 5
the Anthropocene 60, 122
aphids 125
Apollo (Greek god) 18
architecture
   and cows 170, 174, 176
   and trees 118–19, 192
art, rats in 21–2
arum lilies 116

Ashe, E. Wellesley 62-3
Assyrian art
　pollination in 132
Australia
　wet world folklore 154
Australian Aboriginal stories
　the bunyip 154
　Dreamtime story of the Firehawk 72-4
Aztecs 116

Babylon 156
badgers 106, 205-6
Banksy
　Crude Oils exhibition 21-2
baobab trees 118-19
Barnett, Henrietta 43-5
Batiscombe, Mr (forest officer) 184-5
Bear (dog) 20-1
beavers 181
beekeepers 124
bees 124, 132, 138-45
　and butterflies 140
　and guerilla gardening 131
　in folklore 143-5
　honeybees 140
　Oxford Plan Bee project 133-4
　in Singapore 135
　and wasps 139-42
beetles 125
Berdoy, Manuel 16-17
Bethnal Green
　and Fanny Wilkinson 42
　cow-keeping 175, 179
　Nature Reserve 108-11
Biblical stories
　Daniel in the lion's den 51-2
Biko, Steve 112-13
biodiversity

　in cities 108-11, 122, 135
　　Oxford 134, 142
　　Singapore 134-6
　and cows 172, 181
　loss 122, 126, 158
　in the South African Cape Flats 114
　in urban spaces 108-11
　and wetlands 149-50, 165
birds 68, 69-85
　bird corridors 161
　black kites (hawks) 70-3, 80
　black-headed herons 47
　blackbirds 106
　decline in diversity 128
　the Firehawk story 72-3
　guinea fowl 69-71
　hadada ibis 79-83
　the honeyguide bird 143
　kori bustards 47
　maribou storks 74-5
　migratory birds and wetlands 150, 160, 161
　in Nairobi National Park 5
　pigeons 10-13, 17, 23-7, 72
　vultures 74-6
Black Death 19-20
Black, Jack 21
black kites (hawks) 70-3, 80
blackbirds 106
Blixen, Karen 1
bogs 155
bomb sites
　making gardens from 108-11, 121
books
　and Kenyan nature 78
Boston (United States)
　urban dairying 173
brewers in London 175

INDEX

Britain
  allotments 98–103
  guerilla gardens 98
  insect decline 125
  Kew Gardens 187
  land enclosure 34–7, 45–6
  wetlands
    River Mersey 160–1
    Sussex 161–2
  Winstanley and the Diggers 33–4, 46
  see also London
British Empire 6–7
British folklore
  and wetlands 154
bubonic plague 19–20
Bugs Matter survey 125
Bulgaria
  indigenous lions in 54, 55
Burrows, Ruth 94
Butler, Octavia 103
butterflies 140

Cairo 157
Cambridge
  cows in 172–3
Cameroon
  Douala 199
canals 166
carbon storage in wetlands 149
caterpillars 125, 126, 141, 142, 145
cats 13–14, 15, 17
cattle
  legend of the Ngong Hills 2
  Luo peoples and cattle-keeping 170–1
  in Maasai culture 2, 5, 57–9
  rewilding experiment in the Netherlands 127–8
  see also cows

Celliers, Jacob Daniël 188
Celtic mythology
  telling the bees 144
chamomile 111
charcoal 193–4
Charles III, King 144
Chelsea Flower Show 120
chemical weapons and pesticides 126
Cher Ami (homing pigeon) 24
children
  benefits of gardening for 89
  and cows in London 177
Churchill, Winston 62, 63
cicadas 204
Cini, Dr Alessandro 142
cities 204–5
  bees in 132, 138–45
  cows in 169–70, 170, 172–82
  as ecosystems 7
  Mesopotamia 156
  plants in 137–8
  rewilding in 136–7
  Seoul 195, 196, 197
  Sheffield 194–5, 197
  Singapore 134–6, 195, 196
  supporting insects 130–42
  trees in 182, 186–90, 194–6, 198–202
  Venice 191–2
  wetland cities 155, 156, 157–8, 159–61, 166
  see also London; Nairobi; urban spaces
Clark, James D. 188
climate crisis 159, 181
coastal erosion 157
cognitive sanctuaries 32
colonialism
  big-game hunting 62–4

233

colonialism – *cont'd.*
  and Kikũyũ communities 77
  and the Maasai peoples 60–2
  and Nairobi National Park 64
  and the Nubian community 95
  and trees 184, 186–8, 189
  and wetlands 150
  see also European settlers
community gardens 103–4
  in Singapore 135
concrete, reinforced 138
Congo Basin forest management 192–3
COVID-19 pandemic
  and allotments 99, 100, 101, 121
  lockdowns 105–6
Cowie, Mervyn 64
cows 167–82
  calves 169
  in cities 169–70, 170, 172–82
  drovers 173–4
  future of 181–2
  and high-tech industrial farming 181
  and human evolution 171
  and regenerative agriculture 181
  Rosetta (family cow) 167–9, 173, 182
  as symbols of conflict 181
  see also cattle; milk
culture-nature divide
  and trees 184, 186–7
cuneiform writing 156

dahlias 115–17
Daniel, Biblical character 51–2
Daniels, Phil 25
Darley, Gillian 40
death
  bodies in marshland 150
  telling the bees 144

Delhi 157
Denmark 125
the Diggers 33–4
dogs 105, 106
  rat-catching 20–1
drovers 173–4

Earth Trust 151, 165
ecological grief 197
ecological memory 158
Edward II, King of England 174
Egypt
  Aswan High Dam 157
  bees in Ancient Egypt 143
  Heqet (wetland goddess) 154
  the ibis in mythology 82
  the Nile 148, 152, 154, 156–7
elephant grass (*mabingobingo grass*) 3–4
elephants 181
Eliot, Charles 92
Elizabeth II, Queen
  death of 144
Ellis, Dr William 117, 118
Engai, Maasai god 58
Eridu, Mesopotamia 156
estuaries 155
estuarine pools 161
Euphrates, River 155, 157
Euripides 54
European settlers
  in Nairobi 61–2

farms
  and dairies in London 176–7
  Nairobi micro-farms 90
  Oak Grove farm, Boston 173
  organic farm in Kibera 92–8
Fawcett, Millicent Garrett 40

INDEX

fear
  and coexstence with lions 52-3, 54-6, 59-60
  and folklore about wetlands 153-5
feminism 40
fens 161
fig wasps 142
films
  mice and rats in 17-18, 22
Finland
  wet world folklore 154
Firehawk story 72-4
First World War
  allotments 100-1, 133
    the 1918 Allotment project 100-3, 107, 126
  homing pigeons 24
  poetry and gardening literature 102
  and tear gas 126
fleas 19
flies 141, 142
floodplains 155
Florida Everglades 162-3, 165
flowers
  in allotments 104, 132
  in cities 131, 139
  dahlias 115-17
  and guerrilla gardeners 131
  hollyhocks 131
  orchids 142
  pollination 132, 133
  poppies 107-8
  sunflowers 95-6
flu pandemic (1918-19) 100, 103
folklore
  bees in 143-5
  the Firehawk story 72-4
  lions in 49-50
  of the Maasai 2-3, 57-8
  Native American myth of the ibis 82-3
  parable of the hummingbird 84-5
  wetlands in 150-1, 152-5
  see also Greek mythology
food production
  future of 181-2
  milk and urban cow-keeping 173, 175, 176-80
  see also agriculture; allotments
food scarcity
  and rewilding 128
food security 34
food webs
  and insecticides 125
forests 184, 190-7
  in cities 194-6, 200-2
  and ecological therapy 197
  of the future 202
  human use of 193-4
  trees and music 198-202
  and Venice 191-2
  see also trees
foxes 17, 105-6
France
  First World War and homing pigeons 24
Freeman's Wood, Lancaster 36-7
freesias 116
Friess, Dan 163, 164
frogs 136, 204
fruit
  and pollinating insects 132

Ganges, River 161
Garden City movement 45
garden tubs/planters 138

gardening
  allotments 98–103
  co-creating gardens with nature 108
  community gardens 103–4
  and the COVID-19 pandemic 100
  guerilla gardening 120
  health benefits of 89
  organic farm in Kibera 92–8
  secret gardens 87–9, 90
  urban gardening in Nairobi 91
gardens 29–46
  Bethnal Green Nature Reserve 108–11
  Grandin's *Le Jardin sonore* 199
  Phytology 110
  remembrance gardens 119–20
  University of the Western Cape food garden 112–17
  Victorian London 37–45
  Zen rock garden 31–3, 45
Georgian architecture
  and cows in London 175–6
German Museum of Masterpieces of Science and Technology 138
Germany
  insect decline 125
Ghar el Melh 158, 159–60
  giraffes 5
global warming
  and wetlands 149
gold extraction
  Johannesburg 157–8
Grandin, Lucas
  Le Jardin sonore 199
Greek mythology
  and the ibis 82
  the lion in 52–4, 55–6, 67
  mice in 18

see also folklore
Green Belt Movement 84, 196
Grenfell Garden of Peace 119–20
Grow2Know 120
Guatemala 116
guerilla gardeners 98, 131–2
guinea fowl 69–71
Gujjar people (South Asian pastoralist group) 171
Gunner, Gary 47

habitat loss
  and insect decline 127
hadada ibis 79–83
Hampstead Garden Suburb 43–4
hares in cities 203, 206
hawks (black kites) 70–3
Hayden-Smith, Tayshan 119–20
health benefits
  of gardening 89
hedgehogs 106
Heqet, Egyptian river goddess 154
Hercules and the Nemean lion 53–4, 55–6
Hermes, Greek messenger god 82
Highsmith, Tod 154
Hill, Octavia 37–9, 40, 43, 45, 46
  Red Cross Garden, Southwark 40–2
Hindus
  British Hindus and the Mersey 161
honeybees 76
Hopscotch (cat) 13–14, 15
hornets 140
horses
  Kenya 6
  rewilding experiment in the Netherlands 127–8

INDEX

housing in Victorian London 37–8
    Hampstead Garden Suburb 43–4
    Red Cross Cottages 41
hoverflies 126
hunting
    big-game hunting in colonial Kenya 62–3
hyenas
    in folklore 49–50, 153

ibises 79–83
The Iliad 18
India
    Maldhari pastoralists and Asiatic lions 65–6
    New Delhi 169
Industrial Revolution 160
insects 123–45
    and the 1918 Allotment 102
    in cities 130–42
    decline 124–7
        and pesticides 125–6
        the 'windshield phenomenon' 125
    pollinators 102, 131–4, 135, 204
    stinging insects 140–1
    wasps 139–42
    and wetlands 150
    see also bees
Iran 155
Iraq 155

jacaranda trees 186–90, 201
    blossoms 186
    seed pods 186
jackals 49–50
Jacobs, Rosemary 113
Jaftha, Malcolm 116–17
Jaftha, Moses 116

Jaftha's Flower Farm 116–17
Jameson, Frank Walter 188
Janney, Christopher
    Sonic Forest 198
Japan 29–33
    Kyotot rock garden 31–3, 45
    wet world folklore 154
Johannesburg 157–8, 186

Kahneman, Daniel 141
Kalahari San peoples
    bees and the creation story 143
Kamunyak, lioness 51, 52, 67
Karura Forest 84, 196, 197
Kent Wildlife Trust 125
Kenya
    Athi River 5
    colonial administration and the Maasai peoples 60–2
    Eldoret 188
    European settlers 5
    Forest Service 191
    Green Belt Movement 84, 196
    jacaranda trees 187–8, 188–90, 201
    Kibos 191, 195, 197
    Lake Victoria (Nam Lolwe) 147, 152, 163, 164
    Nakuru 188
    Ngong Forest 123
    Ngong Hills 2
    political upheaval (1982) 90–1
    post-election violence (2008) 91–2
    Samburu National Reserve 51
    Samia community 4
    Thimlich Ohinga (Luo settlement) 170, 171–2
    White Highlands 187
    Wildlife Service 62

see also Luo peoples; Maasai people; Nairobi; Nairobi National Park
Kibaki, Mwai 92
Kikuyu grass 4
Kikũyũ people 5, 77
Kilmer, Joyce
   'Trees' 197–8
Kitengela
   Maasi people in 57
Kosano 147–8, 150
kraals and cattle keeping 170–1
Kyrle Society 39

lacewings 125
ladybirds 125
land
   conservation 37
   enclosures 34–7, 45–6
   memory in 97
   organic farming in Kibera 92–8
   projects with 190–1
   redevelopment 36–7
   see also agriculture; forests
leopards 123–4
lice 19
lions
   adopting other baby animals 51
   Asiatic lions 65–6
   Biblical story of Daniel 51–2
   fear and coexistence with lions 52–3, 54–6, 65–8
   folk tales of 49–50
   Greek mythology featuring 52–4, 55–6, 67
   Kamunyak 51, 52, 67
   Lion Lights 59–60, 65
   and the Maasai peoples 56–7, 59–60
   in Nairobi National Park 4, 47–8, 50, 62, 68
Lisbon 186, 190
Liverpool 160–1
Livingston, Ken 24, 25–6
London 6–7, 157
   Arsenal's Emirates Stadium 193
   Chiswick 172
   cows in 172, 174–80
   and mews dairies 176–9
   crossing-sweepers 175
   Georgian architecture 175–6
   Grenfell Tower fire 119–20
   Hampstead Garden Suburb 43–4
   Hope Gardens 120
   Islington 174
   Kew Gardens 189
   Metropolitan Public Gardens Association (MPGA) 39
   nature in 7
   pigeons in 24–6
   Richmond Park 172
   Smithfield Market 174
   the Strand 25
   Trafalgar Square 25
   Victorian women and green space development 37–45
   Whitechapel 175
   Wimbledon Common 172
   see also Bethnal Green
Los Angeles 186, 190
Ludwig, Baron Carl Ferdinand Heinrich von 187
Luo peoples 4
   cattle keeping and kraals 170–1
   and wetlands 148, 150, 151–3, 158, 163
Lutyens, Edwin 44

INDEX

Maasai peoples 2, 5, 56–60
   cattle and Maasai culture 2, 5, 57–9, 65
   colonial administration and the
      Maasai Agreements 60–2
   and European settlers 5
   lactase persistence 171
   and lions 56–7
      initiation rites 59–60
   and Nairobi 57
   and *rungus* 77
Maathai, Professor Wangari 83–4, 85, 196
mabingobingo grass 3–4, 78
McDonald's 185
McPherson, Sandra
   'Lions' 67–8
Mad-Ha Ranwasii 164
Makan, Azzara 115, 116
Mandela, Nelson 112–13
Mandela, Winnie 112
mangroves 149, 155, 164, 165, 199
maribou storks 74–5
marshes 149, 150, 154, 155, 156, 158, 165
   tidal 160–1
marshmallow 111
Martin, Remy 1
Mary Poppins film 25
Meath Gardens, Bethnal Green 42
medicinal plants 110–11, 193–4

memory
   and organic farming in Kibera 92–8
   trees and cultural memory 117–19
mental health
   benefits of gardening for 89
   benefits of urban forests for 196–7
Mersey, River 160–1
Mesopotamia (Oxford) 155
Mespotamia (Iraq) 155, 157

Marsh Arabs 158
Mexico 116
Mexico City 186
Miccosukee peoples
   and the Florida Everglades 162–3
mice 14–15, 17–18, 19
milk
   dairies in London 176–9
   humans and lactase persistence 171
   outsourcing production of 181
   pasteurisation 177, 179
Moha, Youth Reform member 95, 96
Moi, Daniel arap 90
mole-catching 76–7, 78, 79
Molorchus and Hercules 53, 54
mongoose 129
Monier, Joseph 138
mosquitos 7, 148
moths 133, 142
Mozambique 66
MPGA (Metropolitan Public Gardens
      Association) 39
mudbrick technology 156
music and trees 198–202
   musical instruments 200–1
   'Plant an Orchestra' project 201
Myatt's Fields, Camberwell 42
myths *see* folklore; Greek mythology

Naga Cork 73
Nairobi 4–6
   birds 72, 74–5, 79–81
   climate 79–80
   cows 169
   curfews (1982) 90–1
   Kamukunji Park 83
   Karen neighbourhood 1–3, 6, 38, 76,
      80, 92

Nairobi – *cont'd.*
  Karura Forest 84, 196, 197
  Kibera 62, 92
    organic farm 92–8
  Lang'ata 62
  and Maasai peoples 57
  micro-farms 90
  racial segregation and land
    distribution 92
  trees 83–4, 186
    jacaranda 188–90, 201
    Nairobi Arboretum 183–5, 186, 189–90, 191, 192, 201
  University 97
Nairobi National Park 4–5, 60, 64–5
  creation of 61
  lions in 4, 47–8, 50, 62, 68
Nam Lolwe (Lake Victoria) 147, 148, 152, 163, 164
Native American myths
  and the ibis 82–3
nature-culture divide
  and trees 184, 186–7
Nemean lion legend 53–4
Netherlands
  dahlias 116
  Eindhoven 22–3
  rewilding experiment 127–8
New Delhi 75, 169, 186
New York 136, 157
  rats 20
  Roosevelt Island 'healing forest' 201
  Sheffield Farms in Brooklyn 180
  Sonic Forest installation 198
New Zealand
  kauri tree and cultural memory 117–18
Nile River 148, 154

flooding 156–7
the White Nile and the Blue Nile 152
Norse peoples
  lactase persistence 171
Nubian families
  and the Kibera organic farm 95, 117

Odinga, Raila 92
Ogada, Curity 74–5
Ogada, Mordecai 163
Okeechobee, Lake 162
Oloo (gardener) 3, 11–12, 87, 91, 168
orchids 142
organic farming
  Kibera 92–8
Oriel, Elizabeth 184, 186, 189
oryxes and lions 51
Òsanyìn, Yoruba deity 111
Out of Africa 1
Oxford 203, 205–6
  community garden 103–4
  garden water features 136
  guerilla gardeners 98, 131–2, 139
  Mesopotamia (river island) 155–6
  Oxford Council Pollinator Action Plan 133
  pollinating insects 131–2, 133–4, 138–9
  Port Meadow 172
  urban allotments 101, 104, 137
Oxford University
  Oxford Plan Bee project 133–4, 142

papyrus 152, 166
Paris 157
parks
  in Victorian London 38, 39, 40, 42–3

INDEX

Paul, allotmenteer 137, 138
peatlands 149
pest control and ibises 81
pesticides 102
    and the 1918 Allotment project 126
    and cities 138-9
    and insect decline 125-6, 127
Philadelphia
    Awbury Arboretum 199-200
Phytology 110
phytoremediators 95-6
pigeons 10-13, 17, 23-6, 27
    homing pigeons in wartime 23-4
    in London 24-6
    Tony's pigeon 10-11, 12, 23, 27
plants
    in cities 137-8
    medicinal 110-11, 193-4
    organic farm in Kibera 95-6
    phytoremediators 95-6, 97
    in remembrance gardens 119
    resilience of 121-2
    ruderal species 107-8, 111, 121
    sacrificial crops 127
    the toothbrush bush 193
    the vegetal effect of 186
    as witnesses 117-19, 121
    see also flowers; gardens; trees; vegetables
poetry
    and the 1918 Allotment 102, 103
Poland
    Łódź and the Sokołowka River 159
Polgreen, Dr Kim 133
pollinating insects 102, 131-4, 135, 204
poppies 107-8
Pretoria
    jacaranda trees 188, 189, 190

pythons 4

railways
    and cows in London 178
    and Nairobi 185
    railway tracks and allotments 98-9
Ramsar Wetland City 160
Ratatouille (film) 22
rats 14-23
    albino pups 14-15
    in art 21-2
    and the bubonic plague 19-20
    catching 20-1
    laboratory rats 15, 16, 17
    in urban spaces 15-16, 17
    in Victorian England 21
Rebecca (forester) 192, 194
Red Cross Garden, Southwark 40-2
Redford, Robert 1
reinforced concrete 138
representative heuristic 141
rewilding 127-31, 145, 161, 202
    in cities 136-7
    pocket rewilding 133-4
rhinoceroses 4
ribwort plantain 110, 111
rivers 62, 72, 149, 152, 154, 155, 158, 159, 160-1, 164, 181, 199
Roosevelt, Theodore 62-3
ruderal plants 107-8, 111
rural areas
    rats in 20
rural workers
    drovers 173-4
Ruskin, John 37
Russell, Edmund P. 126-7
Rusucmona 159

INDEX

Samburu National Reserve 51
Samburu peoples 51, 77
San Francisco 9–10
Sanders, T W
   Kitchen Garden and Allotment 101, 102
Scandinavia
   Norse peoples and lactase persistence 171
scavenging birds 71–2, 73–6
Schama, Simon
   Landscape and Memory 137
Scotland 194
Second World War
   allotments 101, 102
   former bomb sites 109–10
secret gardens 87–9
seeds
   the 1918 Allotment 101–2
   sowing 87–8, 89–90
   swapping 112, 115
Seminole peoples
   and the Florida Everglades 162–3
Seoul Forest 195, 197
sheep 49–50
   driven into London 174
Sheffield 194–5, 197
Singapore 134–6, 195
Sislu, Walter 112–13
Skinner, Sam 99, 102
Sloan, Kevin 136
Smithsonian Institution 63
Smythe, Michael 110–11
snakes 3–4
snapdragons 116
soil
   de-toxifying 94–6
   microbes in 89
South Africa 92

Anglo-Boer War 187
apartheid
   colonial settlement 114–15
   and freedom fighters 112–13
Cape Town 187
dahlias 115–17
hadada ibis 81
jacaranda trees 187, 188, 189
Jaftha's Flower Farm 116–17
Johannesburg 157–8
Kimberley 188
Pretoria 188, 189, 190
see also UWC (University of the Western Cape)
South America 130
Southgate, Fran 161–2, 165
Spanish flu pandemic (1918-19) 100, 103
Speke and Garston Coastal Reserve 160–1
spiders 141
Sprengel, Christian Konrad
   on insects and pollination 132–3
squirrels 19
stags in cities 203, 206
Streep, Meryl 1
Sullivan, Robert 20
Sumerians 156
Sun Ra Arkestra 200
sunflowers 95–6
Sussex Wildlife Trust 161–2, 165
swamps 149, 155
Syria 155

The Tail of Despereaux 17–18
Tambo, Oliver 112–13
tear gas 126
Tempelman, travelling nurseryman 188
Thames, River 151, 165

Thimlich Ohinga (Luo settlement)
   cattle 170, 171-2
Thoth, Egyptian god 82
Tigris, River 155, 157
Tiryns 54
Toda people (South Asian pastoralist
   group) 171
Tom and Jerry 17
Tony's pigeon 10-11, 12, 23, 27
Torres Strait Islanders 72
travelling
   trees and travels 185-6
trees 105
   and the boundaries between nature
      and culture 183-4
   forests 184, 190-7
   indigenous and exotic 183-4, 185
   jacarandas 186-90, 201
   Kilmer's poem 'Trees' 197-8
   and maribou storks 74
   and music 198-202
   in Nairobi
      Nairobi Artboretum 183-5, 186
      planting 83-4
   nature and culture 184, 186-7, 197
   in Sheffield 194-5
   and urban spaces 182
   as witnesses 117-19, 121
   see also forests
Tunisia
   Ghar el Melh 158, 159-60
Turere, Richard 57, 58-9, 65
Turkey 155
   temple ruins in Gülpinar 18
Tversky, Amos 141

Uganda
   Banyankole peoples and cattle 182

Umpierre, Diana 162-3, 165
United Kingdom *see* Britain
United States
   Awbury Arboretum, Philadelphia
      199-200
   Florida Everglades 162-3, 165
   Smithsonian Institution 63
   urban cow-keeping 180
   urban dairying 173
   see also New York
Unwin, Raymond 44
urban spaces 8, 9-27, 122
   biodiversity in 108-11
   birds in
      ibises 80-2, 83
      pigeons 24-6, 27
   ecosystems and waste management
      74-6
   gentrification of 45
   greening of 22-3
      in Victorian London 37-45
   rats in 15-16, 17, 20, 22-3, 26-7
   and rural enclosure 36-7
   see also cities
Uruk, city-state 156
UWC (University of the Western
   Cape)
   and apartheid resistance 114-15
   food garden 112, 113-14, 115-17
   collaborators 113-14
   seed swaps 112, 115

Vauxhall Park, London 40, 42-3
vegetables
   the 1918 Allotment 101-2
   organic farming in Kibera 96
   and pollinating insects 132
Venice 191-2

INDEX

Victoria, Lake (Nam Lolwe) 147, 148, 152, 163, 164
Victoria, Queen 21
Victorian England
    rats in 21
Victorian women
    London and green space development 37–45
visitors' stories
    and the 1918 Allotment 102–3
vultures 75–6

Warũingi, Mzee 76–9, 80, 81, 85
wasps 139–42
    attitudes to 139–40, 145
    as hunters 141
    as pollinators 141–2
    species of 140
water management 157
water purifiers, wetlands as 149
Watsonia 116
Welford Dairy, London 177–9, 180
Welsh farmers 96–7, 176, 177
West Africa
    Yoruba peoples 111
wetlands 145, 147–66
    benefits of 149–50
    and cities 155, 156, 157–8, 159–61
    communities and wetland projects 165
    development of 164–5
    flooding 149
    in folklore 150–1, 152–5

    and Luo peoples 148, 150, 151–3
    mangroves 149, 155, 164, 165
    marshes 149, 150, 154, 155
    moving water 151
    Nam Lolwe (Lake Victoria) 148, 152
    the Nile River 148
    seasonal floods 157, 158
    shaping 165–6
    the Thames 151
Whittlesey, Major Charles 24
Wild Oxfordshire 144
Wilkinson, Fanny 39, 40, 42–3
will-o'-the-wisps 150
Winstanley, Gerrard 33–4, 46
women
    Victorian reformers and green spaces 37–45
Wordsworth, William 164
worms 106
Wye, River 164

yarrow 110, 111
yellowjackets 140
Yoruba peoples 111
Youth Reform Group
    organic farming in Kibera 93–8

zebras 5
Zeus 53, 54
Zimbabwean folklore
    honeyguide bird 143

Acknowledgements

Nyasaye & the ancestors – always.

With enduring appreciation to my beloved husband Peter, who actively supports my dream-chasing.

To Sandra McPherson's family for their kind permission to include her poem 'Lions' and to Sister Teresa for the kind permission to include Ruth Burrow's poem 'I Made a Garden for God'.

I have much to be grateful for, thanks to many. To all those named in these pages who revealed just how magical life in the city can be. To my agent, Caro Clarke, who keeps opening doors. The team at Octopus Books, including Jessica Minocha, who taught me how to feel rhythm and pacing. All the people who helped to bring this book to life including Rachel Silverlight, Susanne Hillen, Mel Four and Sarah Parry. To my wonderful first readers – Terri Mulholland, Lynn Taylor, Ollie Randall – who gently nudge me in the right direction. To Alex Braslavsky, ever inspiring writing buddy. To Sam Skinner and Julia Utreras, for the gift of collaboration. To Professor Ciraj Rassool for the opportunity to be involved with a garden in Cape Town. To Amy Rowland, Earth Trust and all those who care for a special part of Oxfordshire where a piece of my heart will always live.

My heartfelt thanks to Johanna Zetterström-Sharp, dear friend and collaborator whose countless conversations helped me turn scattered thoughts into something worth putting on the page. I owe chapter nine to our work together on milk. That chapter would also have not been possible without the support of the Arts and Humanities Research Council for our project: Milking it: Colonialism, Heritage and Everyday Engagement with Dairy.

Erokamano ahinya to my parents who gifted me a childhood that offered the world.

And to B, who would not read one word until this was published: I hope it was worth the wait.

Shukra! Skukran! Shukrani!

About the Author

JC NIALA is an award-winning writer, environmental historian and anthropologist specializing in human-nature interactions. JC works as the Deputy Director and Head of Research, Teaching and Collections at the History of Science Museum, University of Oxford. She has been on BBC's 'The Conversation' and 'Thinking Allowed', and she wrote and narrated the nature podcast series 'The Root of the Matter' with Wellcome Collection. She has appeared on panels at The Garden Museum, London and The Old Fire Station in Oxford, and has participated in the Edinburgh International Book Festival and Cheltenham Science Festival, as well as the Hay Literary Festival in the UK and Kenya. JC lives in Oxford and Nairobi – and has lived in other cities including London and San Francisco.

RAISING READERS
Books Build Bright Futures

Dear Reader,

We'd love your attention for one more page to tell you about the crisis in children's reading, and what we can all do.

Studies have shown that reading for fun is the **single biggest predictor of a child's future life chances** – more than family circumstance, parents' educational background or income. It improves academic results, mental health, wealth, communication skills, ambition and happiness.[1]

The number of children reading for fun is in rapid decline. Young people have a lot of competition for their time. In 2024, 1 in 10 children and young people in the UK aged 5 to 18 did not own a single book at home.[2]

Hachette works extensively with schools, libraries and literacy charities, but here are some ways we can all raise more readers:

- Reading to children for just 10 minutes a day makes a difference
- Don't give up if children aren't regular readers – there will be books for them!
- Visit bookshops and libraries to get recommendations
- Encourage them to listen to audiobooks
- Support school libraries
- Give books as gifts

There's a lot more information about how to encourage children to read on our website: **www.RaisingReaders.co.uk**

Thank you for reading.

---

[1] OECD, '21st-Century Readers: Developing Literacy Skills in a Digital World', 2021, https://www.oecd.org/en/publications/21st-century-readers_a83d84cb-en.html

[2] National Literacy Trust, 'Book Ownership in 2024', November 2024, https://literacytrust.org.uk/research-services/research-reports/book-ownership-in-2024